Common Core Subject Test Mathematics Grade 6

Student Practice Workbook

+ Two Full-Length Common Core Math Tests

Math Notion

www.MathNotion.com

Common Core Subject Test Mathematics Grade 6

Common Core Subject Test Mathematics Grade 6

Common Core Subject Test Mathematics Grade 6

Published in the United State of America By

The Math Notion

Web: WWW.MathNotion.com

Email: info@Mathnotion.com

Copyright © 2021 by the Math Notion. All rights reserved. No part of this publication may be reproduced, stored in a retrieval system, or transmitted in any form or by any means, electronic, mechanical, photocopying, recording, scanning, or otherwise, except as permitted under Section 107 or 108 of the 1976 United States Copyright Ac, without permission of the author.

All inquiries should be addressed to the Math Notion.

ISBN: 978-1-63620-072-9

Common Core Subject Test Mathematics Grade 6

The Math Notion

Michael Smith has been a math instructor for over a decade now. He launched the Math Notion. Since 2006, we have devoted our time to both teaching and developing exceptional math learning materials. As a test prep company, we have worked with thousands of students. We have used the feedback of our students to develop a unique study program that can be used by students to drastically improve their math scores fast and effectively. We have more than a thousand Math learning books including:

– SAT Math Prep

– ACT Math Prep

– SSAT/ISEE Math Prep

– Accuplacer Math Prep

– Common Core Math Prep

–many Math Education Workbooks, Study Guides, Practice and Exercise Books

As an experienced Math test preparation company, we have helped many students raise their standardized test scores—and attend the colleges of their dreams: We tutor online and in person, we teach students in large groups, and we provide training materials and textbooks through our website and through Amazon.

You can contact us via email at:

info@Mathnotion.com

Common Core Subject Test Mathematics Grade 6

Get the Targeted Practice You Need to Ace the Common Core Math Test!

Common Core Subject Test Mathematics Grade 6 includes easy-to-follow instructions, helpful examples, and plenty of math practice problems to assist students to master each concept, brush up their problem-solving skills, and create confidence.

The Common Core math practice book provides numerous opportunities to evaluate basic skills along with abundant remediation and intervention activities. It is a skill that permits you to quickly master intricate information and produce better leads in less time.

Students can boost their test-taking skills by taking the book's two practice Common Core Math exams. All test questions answered and explained in detail.

Important Features of the 6th grade Common Core Math Book:

- A **complete review** of Common Core math test topics,
- Over 2,500 practice problems covering all topics tested,
- The most important concepts you need to know,
- Clear and concise, easy-to-follow sections,
- Well designed for enhanced learning and interest,
- Hands-on experience with all question types,
- **2 full-length practice tests** with detailed answer explanations,
- Cost-Effective Pricing,

Powerful math exercises to help you avoid traps and pacing yourself to beat the Common Core test. Students will gain valuable experience and raise their confidence by taking 6th grade math practice tests, learning about test structure, and gaining a deeper understanding of what is tested on the Common Core math grade 6. If ever there was a book to respond to the pressure to increase students' test scores, this is it.

Common Core Subject Test Mathematics Grade 6

WWW.MathNotion.COM

... So Much More Online!

- ✓ FREE Math Lessons
- ✓ More Math Learning Books!
- ✓ Mathematics Worksheets
- ✓ Online Math Tutors

For a PDF Version of This Book

Please Visit WWW.MathNotion.com

Common Core Subject Test Mathematics Grade 6

Contents

Chapter 1 : Review of the Whole Number Operations..................11
   Adding Whole Numbers 12
   Subtracting Whole Numbers 13
   Multiplying Whole Numbers..................14
   Dividing Hundreds..................15
   Long Division by Two Digits 16
   Division with Remainders 16
   Rounding Whole Numbers 17
   Whole Number Estimation 18
   Answers of Worksheets 19

Chapter 2 : Integers and Number Theory..................21
   Adding and Subtracting Integers 22
   Multiplying and Dividing Integers 23
   Order of Operations 24
   Ordering Integers and Numbers..................25
   Integers and Absolute Value 26
   Factoring Numbers 27
   Prime Factorization..................27
   Divisibility Rules..................28
   Greatest Common Factor 29
   Least Common Multiple 30
   Answers of Worksheets 31

Chapter 3 : Fractions 35
   Simplifying Fractions..................36
   Adding and Subtracting Fractions 37
   Multiplying and Dividing Fractions 38
   Adding and Subtracting Mixed Numbers 39
   Multiplying and Dividing Mixed Numbers 40
   Answers of Worksheets 41

Chapter 4 : Decimals..................43
   Adding and Subtracting Decimals..................44
   Multiplying and Dividing Decimals 45
   Comparing Decimals..................46

WWW.MathNotion.Com

Common Core Subject Test Mathematics Grade 6

Rounding Decimals 47
Convert Fraction to Decimal 48
Convert Decimal to Percent 49
Convert Fraction to Percent 50
Answers of Worksheets 51

Chapter 5 : Proportions, Ratios, and Percent 53
Simplifying Ratios 54
Proportional Ratios 55
Similarity and Ratios 56
Ratio and Rates Word Problems 57
Percentage Calculations 58
Percent Problems 59
Discount, Tax and Tip 60
Answers of Worksheets 61

Chapter 6 : Exponents and Radicals Expressions 63
Adding and Subtracting Exponents 64
Multiplication Property of Exponents 65
Zero and Negative Exponents 66
Division Property of Exponents 67
Powers of Products and Quotients 68
Negative Exponents and Negative Bases 69
Scientific Notation 70
Square Roots 71
Answers of Worksheets 72

Chapter 7 : Measurements 75
Reference Measurement 76
Metric Length Measurement 77
Customary Length Measurement 77
Metric Capacity Measurement 78
Customary Capacity Measurement 78
Metric Weight and Mass Measurement 79
Customary Weight and Mass Measurement 79
Temperature 80
Time 81
Answers of Worksheets 82

Chapter 8 : Algebraic Expressions 84
Find a Rule! 85

WWW.MathNotion.Com

Common Core Subject Test Mathematics Grade 6

Translate Phrases into an Algebraic Statement 86
Simplifying Variable Expressions 87
The Distributive Property 88
Evaluating One Variable Expressions 89
Combining like Terms 90
Answers of Worksheets 91

Chapter 9 : Equations and Inequalities 93
One–Step Equations 94
One–Step Equation Word Problems 95
Two-Steps Equations 96
Multi–Step Equations 97
One-Step Inequalities 98
Graphing Inequalities 99
Two-Steps Inequality 100
Multi-Step Inequalities 101
Answers of Worksheets 102

Chapter 10 : Geometry and Solid Figures 105
Angles 106
Pythagorean Relationship 107
Triangles 108
Polygons 109
Trapezoids 110
Circles 111
Cubes 112
Rectangular Prism 113
Cylinder 114
Answers of Worksheets 115

Chapter 11 : Statistics and Probability 117
Mean and Median 118
Mode and Range 119
Times Series 120
Stem–and–Leaf Plot 121
Quartile of a Data Set 122
Box and Whisker Plots 122
Pie Graph 123
Probability Problems 124
Answers of Worksheets 125

WWW.MathNotion.Com

Common Core Subject Test Mathematics Grade 6

Chapter 12 : Common Core Math Practice Tests 127
   Common Core GRADE 6 MAHEMATICS REFRENCE MATERIALS 129
   Common Core Practice Test 1 131
   Common Core Practice Test 2 147

Chapter 13 : Answers and Explanations 161
   Answer Key 161
   Practice Test 1 163
   Practice Test 2 169

Common Core Subject Test Mathematics Grade 6

Chapter 1 : Review of the Whole Number Operations

Topics that you'll learn in this chapter:

- ✓ Adding Whole Numbers
- ✓ Subtracting Whole Numbers
- ✓ Multiplying Whole Numbers
- ✓ Dividing Hundreds
- ✓ Long Division by One Digit
- ✓ Division with Remainders
- ✓ Rounding Whole Numbers
- ✓ Whole Number Estimation

"Wherever there is number, there is beauty." –Proclus

Common Core Subject Test Mathematics Grade 6

Adding Whole Numbers

✎ **Add.**

1) 5,763
   + 8,238
   _____

2) 6,834
   + 4,998
   _____

3) 3,548
   + 5,693
   _____

4) 2,769
   + 8,872
   _____

5) 3,196
   + 2,936
   _____

6) 7,009
   + 4,992
   _____

✎ **Find the missing numbers.**

7) 3,468 + ___ = 4,102

8) 840 + 2,360 = ___

9) 5,200 + ___ = 7,980

10) 631 + ___ = 2,007

11) ___ + 803 = 3,945

12) ___ + 2,156 = 5,922

13) David sells gems. He finds a diamond in Istanbul and buys it for $4,795. Then, he flies to Cairo and purchases a bigger diamond for the bargain price of $9,633. How much does David spend on the two diamonds? _____

Common Core Subject Test Mathematics Grade 6

Subtracting Whole Numbers

✎ **Subtract.**

1) 10,512
   − 4,411
   _____

2) 5,204
   − 3,679
   _____

3) 8,520
   − 6,483
   _____

4) 8,001
   − 5,224
   _____

5) 11,916
   − 8,711
   _____

6) 5,005
   − 2,008
   _____

✎ **Find the missing number.**

7) 5,263 − ___ = 2,367

8) 7,198 − ___ = 4,742

9) 8,928 − 3,764 = ___

10) 6,511 − ___ = 3,759

11) 7,003 − 5,489 = ___

12) 8,800 − 5,995 = ___

13) Jackson had $7,189 invested in the stock market until he lost $3,793 on those investments. How much money does he have in the stock market now?

_____

Common Core Subject Test Mathematics Grade 6

Multiplying Whole Numbers

✎ **Find the answers.**

1) 2,200 × 31

2) 3,200 × 22

3) 5,790 × 5

4) 5,220 × 3

5) 6,911 × 3

6) 1,998 × 40

7) 2,893 × 5.5

8) 2,254 × 3.5

9) 4,372 × 4.8

10) 3,984 × 2.75

11) 4,900 × 2.5

12) 8,200 × 4.5

WWW.MathNotion.Com

Dividing Hundreds

✎ **Find answers.**

1) 4,440 ÷ 400

2) 1,600 ÷ 40

3) 9,990 ÷ 90

4) 4,200 ÷ 60

5) 6,400 ÷ 8,000

6) 2,700 ÷ 30

7) 3,333 ÷ 30

8) 558 ÷ 45

9) 2,278 ÷ 85

10) 1,683 ÷ 55

11) 1,582 ÷ 35

12) 9,000 ÷ 600

13) 1,000 ÷ 2,500

14) 44.8 ÷ 20

15) 6,800 ÷ 400

16) 1,500 ÷ 5,000

17) 36.60 ÷ 120

18) 7,700 ÷ 700

19) 5,400 ÷ 600

20) 8,000 ÷ 160

21) 18,000 ÷ 9,000

22) 42,000 ÷ 30

23) 480 ÷ 40

24) 63,000 ÷ 900

Common Core Subject Test Mathematics Grade 6

Long Division by Two Digits

✎ Find the quotient.

1) $18\overline{)576}$ 10) $41\overline{)1,476}$

2) $14\overline{)952}$ 11) $53\overline{)2,491}$

3) $21\overline{)588}$ 12) $60\overline{)2,880}$

4) $23\overline{)299}$ 13) $32\overline{)2,912}$

5) $44\overline{)748}$ 14) $77\overline{)8,393}$

6) $26\overline{)234}$ 15) $85\overline{)3,740}$

7) $16\overline{)496}$ 16) $57\overline{)4,617}$

8) $29\overline{)1,479}$ 17) $50\overline{)9,200}$

9) $54\overline{)1,080}$ 18) $25\overline{)15,400}$

Division with Remainders

✎ Find the quotient with remainder.

1) $14\overline{)715}$ 8) $65\overline{)8,624}$

2) $16\overline{)2,750}$ 9) $35\overline{)5,705}$

3) $27\overline{)4,603}$ 10) $92\overline{)13,161}$

4) $58\overline{)2,554}$ 11) $46\overline{)12,214}$

5) $42\overline{)7,732}$ 12) $69\overline{)42,482}$

6) $63\overline{)6,737}$ 13) $85\overline{)6,858}$

7) $71\overline{)9,036}$ 14) $87\overline{)34,304}$

Common Core Subject Test Mathematics Grade 6

Rounding Whole Numbers

✎ Round each number to the underlined place value.

1) 7,<u>5</u>33

2) 9,<u>3</u>74

3) 8,8<u>8</u>3

4) 2,3<u>6</u>8

5) 5,5<u>7</u>7

6) 3,3<u>8</u>1

7) 3,<u>5</u>20

8) 9,3<u>3</u>8

9) 8.<u>5</u>81

10) 33.<u>5</u>7

11) 51.<u>6</u>9

12) 22.<u>1</u>38

13) <u>6</u>,758

14) 11,5<u>5</u>7

15) 8,8<u>3</u>8

16) 5.<u>8</u>89

17) 1.<u>8</u>60

18) 25.<u>0</u>70

19) <u>9</u>.332

20) 49.<u>4</u>8

21) 28.<u>8</u>9

22) 24,3<u>7</u>7

23) 52,1<u>5</u>8

24) 13,8<u>8</u>3

25) 9,<u>6</u>09

26) 17,4<u>5</u>1

27) 18,<u>7</u>68

WWW.MathNotion.Com

Whole Number Estimation

✎ **Estimate the sum by rounding each added to the nearest ten.**

1) 875 + 325

2) 985 + 1,452

3) 2,424 + 4,128

4) 1,576 + 6,279

5) 1,247 + 3,863

6) 6,746 + 5,121

7) 3,924 + 6,456

8) 1,785 + 7,164

9) 1,458
   + 2,442
   ─────

10) 5,689
    + 4,151
    ─────

11) 8,259
    + 4,754
    ─────

12) 6,788
    + 3,954
    ─────

13) 9,123
    + 4,455
    ─────

14) 6,680
    + 5,358
    ─────

15) 3,165
    + 7,124
    ─────

16) 8,859
    + 6,452
    ─────

Common Core Subject Test Mathematics Grade 6

Answers of Worksheets

Adding Whole Numbers

1) 14,001
2) 11,832
3) 9,241
4) 11,641
5) 6,132
6) 12,001
7) 634
8) 3,200
9) 2,780
10) 1,376
11) 3,142
12) 3,766
13) $14,428

Subtracting Whole Numbers

1) 6,101
2) 1,525
3) 2,037
4) 2,777
5) 3,205
6) 2,997
7) 2,896
8) 2,456
9) 5,164
10) 2,752
11) 1,514
12) 2,805
13) 3,396

Multiplying Whole Numbers

1) 68,200
2) 70,400
3) 28,950
4) 15,660
5) 20,733
6) 79,920
7) 15,911.5
8) 7,889
9) 20,985.6
10) 10,956
11) 12,250
12) 36,900

Dividing Hundreds

1) 11.1
2) 40
3) 111
4) 70
5) 0.8
6) 90
7) 111.1
8) 12.4
9) 26.8
10) 30.6
11) 45.2
12) 15
13) 0.4
14) 2.24
15) 17
16) 0.3
17) 0.305
18) 11
19) 9
20) 50
21) 2
22) 1,400
23) 12
24) 70

Long Division by Two Digits

1) 32
2) 68
3) 28
4) 13
5) 17
6) 9
7) 31
8) 51
9) 20
10) 36
11) 47
12) 48
13) 91
14) 109
15) 44
16) 81
17) 184
18) 616

WWW.MathNotion.Com

Common Core Subject Test Mathematics Grade 6

Division with Remainders

1) 51 R1
2) 171 R14
3) 170 R13
4) 44 R2
5) 184 R4
6) 106 R59
7) 127 R19
8) 132 R44
9) 163 R0
10) 143 R5
11) 265 R24
12) 615 R47
13) 80 R58
14) 394 R26

Rounding Whole Numbers

1) 7,500
2) 9,400
3) 8,880
4) 2,370
5) 5,580
6) 3,380
7) 3,500
8) 9,340
9) 8.60
10) 33.60
11) 51.70
12) 22.100
13) 7,000
14) 11,560
15) 8,840
16) 5.900
17) 1.900
18) 25.100
19) 9.000
20) 49.50
21) 28.90
22) 24,380
23) 52,160
24) 13,880
25) 9,600
26) 17,450
27) 18,800

Whole Number Estimation

1) 1,200
2) 2,440
3) 6,550
4) 7,860
5) 5,110
6) 11,870
7) 10,380
8) 8,950
9) 3,900
10) 9,840
11) 13,010
12) 10,740
13) 13,580
14) 12,040
15) 10,290
16) 15,310

Common Core Subject Test Mathematics Grade 6

Chapter 2 :
Integers and Number Theory

Topics that you will practice in this chapter:

- ✓ Adding and Subtracting Integers
- ✓ Multiplying and Dividing Integers
- ✓ Order of Operations
- ✓ Ordering Integers and Numbers
- ✓ Integers and Absolute Value
- ✓ Factoring Numbers
- ✓ Prime Factorization
- ✓ Divisibility Rules
- ✓ Greatest Common Factor (GCF)
- ✓ Least Common Multiple (LCM)

"In order to gain the most, you have to know how to convert Negatives to Positives."
—*Stubborn Clown*

Common Core Subject Test Mathematics Grade 6

Adding and Subtracting Integers

✎ **Find each sum.**

1) $14 + (-6) =$

2) $(-13) + (-20) =$

3) $5 + (-28) =$

4) $50 + (-12) =$

5) $(-7) + (-15) + 3 =$

6) $30 + (-14) + 8 =$

7) $40 + (-10) + (-14) + 17 =$

8) $(-15) + (-20) + 13 + 35 =$

9) $40 + (-20) + (38 - 29) =$

10) $28 + (-12) + (30 - 12) =$

✎ **Find each difference.**

11) $(-18) - (-7) =$

12) $25 - (-14) =$

13) $(-20) - 36 =$

14) $34 - (-19) =$

15) $51 - (30 - 21) =$

16) $17 - (5) - (-24) =$

17) $(35 + 20) - (-46) =$

18) $48 - 16 - (-8) =$

19) $62 - (28 + 17) - (-15) =$

20) $58 - (-23) - (-31) =$

21) $19 - (-8) - (-13) =$

22) $(19 - 24) - (-14) =$

23) $27 - 33 - (-21) =$

24) $58 - (32 + 24) - (-9) =$

25) $36 - (-30) + (-17) =$

26) $27 - (-42) + (-31) =$

WWW.MathNotion.Com

Common Core Subject Test Mathematics Grade 6

Multiplying and Dividing Integers

✎ **Find each product.**

1) $(-9) \times (-5) =$

2) $(-3) \times 9 =$

3) $8 \times (-12) =$

4) $(-7) \times (-20) =$

5) $(-3) \times (-5) \times 6 =$

6) $(14 - 3) \times (-8) =$

7) $12 \times (-9) \times (-3) =$

8) $(140 + 10) \times (-2) =$

9) $10 \times (-12 + 8) \times 3 =$

10) $(-8) \times (-5) \times (-10) =$

✎ **Find each quotient.**

11) $42 \div (-7) =$

12) $(-48) \div (-6) =$

13) $(-40) \div (-8) =$

14) $54 \div (-2) =$

15) $152 \div 19 =$

16) $(-144) \div (-12) =$

17) $180 \div (-10) =$

18) $(-312) \div (-12) =$

19) $221 \div (-13) =$

20) $(-126) \div (6) =$

21) $(-161) \div (-7) =$

22) $-266 \div (-14) =$

23) $(-120) \div (-4) =$

24) $270 \div (-18) =$

25) $(-208) \div (-8) =$

26) $(135) \div (-15) =$

WWW.MathNotion.Com

Common Core Subject Test Mathematics Grade 6

Order of Operations

✎ Evaluate each expression.

1) $7 + (5 \times 4) =$

2) $14 - (3 \times 6) =$

3) $(19 \times 4) + 16 =$

4) $(16 - 7) - (8 \times 2) =$

5) $27 + (18 \div 3) =$

6) $(18 \times 8) \div 6 =$

7) $(32 \div 4) \times (-2) =$

8) $(9 \times 4) + (32 - 18) =$

9) $24 + (4 \times 3) + 7 =$

10) $(36 \times 3) \div (2 + 2) =$

11) $(-7) + (12 \times 3) + 11 =$

12) $(8 \times 5) - (24 \div 6) =$

13) $(7 \times 6 \div 3) - (12 + 9) =$

14) $(13 + 5 - 14) \times 3 - 2 =$

15) $(20 - 14 + 30) \times (64 \div 4) =$

16) $32 + (28 - (36 \div 9)) =$

17) $(7 + 6 - 4 - 7) + (15 \div 5) =$

18) $(85 - 20) + (20 - 18 + 7) =$

19) $(20 \times 2) + (14 \times 3) - 22 =$

20) $18 + 5 - (30 \times 3) + 20 =$

21) $(\frac{7}{5-1}) \times (2 + 6) \times 2$

22) $20 \div (4 - (10 - 8))$

WWW.MathNotion.Com

Common Core Subject Test Mathematics Grade 6

Ordering Integers and Numbers

✏ **Order each set of integers from least to greatest.**

1) $8, -10, -5, -3, 4$ ___, ___, ___, ___, ___, ___

2) $-10, -18, 6, 14, 27$ ___, ___, ___, ___, ___, ___

3) $15, -8, -21, 21, -23$ ___, ___, ___, ___, ___, ___

4) $-14, -40, 23, -12, 47$ ___, ___, ___, ___, ___, ___

5) $59, -54, 32, -57, 36$ ___, ___, ___, ___, ___, ___

6) $68, 26, -19, 47, -34$ ___, ___, ___, ___, ___, ___

✏ **Order each set of integers from greatest to least.**

7) $18, 36, -16, -18, -10$ ___, ___, ___, ___, ___, ___

8) $27, 34, -12, -24, 94$ ___, ___, ___, ___, ___, ___

9) $50, -21, -13, 42, -2$ ___, ___, ___, ___, ___, ___

10) $37, 46, -20, -16, 86$ ___, ___, ___, ___, ___, ___

11) $-18, 88, -26, -59, 75$ ___, ___, ___, ___, ___, ___

12) $-65, -30, -25, 3, 14$ ___, ___, ___, ___, ___, ___

WWW.MathNotion.Com

Common Core Subject Test Mathematics Grade 6

Integers and Absolute Value

✎ **Write absolute value of each number.**

1) $|-2| =$

2) $|-27| =$

3) $|-20| =$

4) $|14| =$

5) $|6| =$

6) $|-55| =$

7) $|16| =$

8) $|2| =$

9) $|54| =$

10) $|-4| =$

11) $|-11|$

12) $|88| =$

13) $|0| =$

14) $|79| =$

15) $|-32| =$

16) $|-17| =$

17) $|42| =$

18) $|-46| =$

19) $|1| =$

20) $|-40| =$

✎ **Evaluate the value.**

21) $|-5| - \frac{|-21|}{7} =$

22) $14 - |3 - 15| - |-4| =$

23) $\frac{|-32|}{4} \times |-4| =$

24) $\frac{|7 \times (-3)|}{7} \times \frac{|-19|}{3} =$

25) $|4 \times (-5)| + \frac{|-40|}{5} =$

26) $\frac{|-45|}{9} \times \frac{|-24|}{12} =$

27) $|-12 + 8| \times \frac{|-7 \times 7|}{7} =$

28) $\frac{|-11 \times 2|}{4} \times |-16| =$

WWW.MathNotion.Com

Common Core Subject Test Mathematics Grade 6

Factoring Numbers

✎ List all positive factors of each number.

1) 12
2) 16
3) 28
4) 34
5) 95
6) 56
7) 65
8) 70
9) 25
10) 48
11) 27
12) 63
13) 72
14) 15
15) 80

✎ List the prime factorization for each number.

16) 10
17) 26
18) 20
19) 30
20) 40
21) 44
22) 55
23) 78
24) 96

Prime Factorization

✎ Factor the following numbers to their prime factors.

1) 6
2) 49
3) 60
4) 4
5) 46
6) 57
7) 54
8) 38
9) 58
10) 62
11) 75
12) 88
13) 93
14) 100
15) 68
16) 90
17) 69
18) 76
19) 86
20) 92
21) 99
22) 77
23) 90
24) 74

WWW.MathNotion.Com

Common Core Subject Test Mathematics Grade 6

Divisibility Rules

✎ **Use the divisibility rules to underline the factors of the number.**

1) 8   2 3 4 5 6 7 8 9 10

2) 18   2 3 4 5 6 7 8 9 10

3) 55   2 3 4 5 6 7 8 9 10

4) 45   2 3 4 5 6 7 8 9 10

5) 20   2 3 4 5 6 7 8 9 10

6) 9   2 3 4 5 6 7 8 9 10

7) 21   2 3 4 5 6 7 8 9 10

8) 28   2 3 4 5 6 7 8 9 10

9) 36   2 3 4 5 6 7 8 9 10

10) 40   2 3 4 5 6 7 8 9 10

11) 39   2 3 4 5 6 7 8 9 10

12) 51   2 3 4 5 6 7 8 9 10

Common Core Subject Test Mathematics Grade 6

Greatest Common Factor

✎ **Find the GCF for each number pair.**

1) 6, 2

2) 4, 5

3) 3, 12

4) 7, 3

5) 5, 10

6) 8, 48

7) 6, 18

8) 9, 15

9) 12, 18

10) 4, 36

11) 6, 10

12) 28, 52

13) 25, 10

14) 22, 24

15) 9, 54

16) 8, 54

17) 42, 14

18) 16, 40

19) 9, 2, 3

20) 5, 15, 10

21) 7, 9, 2

22) 16, 64

23) 30, 48

24) 36, 63

Common Core Subject Test Mathematics Grade 6

Least Common Multiple

✏ **Find the LCM for each number pair.**

1) 6, 9

2) 15, 45

3) 16, 40

4) 12, 36

5) 18, 27

6) 14, 42

7) 6, 30

8) 8, 56

9) 7, 21

10) 8, 20

11) 15, 25

12) 7, 9

13) 4, 11

14) 8, 28

15) 28, 56

16) 40, 50

17) 12, 13

18) 22, 11

19) 36, 20

20) 15, 35

21) 18, 81

22) 30, 54

23) 18, 45

24) 75, 25

Common Core Subject Test Mathematics Grade 6

Answers of Worksheets

Adding and Subtracting Integers

1) 8
2) −33
3) −23
4) 38
5) −19
6) 24
7) 33
8) 13
9) 29
10) 34
11) −11
12) 39
13) −56
14) 53
15) 42
16) 36
17) 101
18) 40
19) 32
20) 112
21) 40
22) 9
23) 15
24) 11
25) 49
26) 38

Multiplying and Dividing Integers

1) 45
2) −27
3) −96
4) 140
5) 90
6) −88
7) 324
8) −300
9) −120
10) −400
11) −6
12) 8
13) 5
14) −27
15) 8
16) 12
17) −18
18) 26
19) −17
20) −21
21) 23
22) 19
23) 30
24) −15
25) 26
26) −9

Order of Operations

1) 27
2) −4
3) 92
4) −7
5) 33
6) 24
7) −16
8) 50
9) 43
10) 27
11) 40
12) 36
13) −7
14) 10
15) 576
16) 56
17) 5
18) 74
19) 60
20) −47
21) 28
22) 10

Ordering Integers and Numbers

1) −10, −5, −3, 4, 8
2) −18, −10, 6, 14, 27
3) −23, −21, −8, 15, 21
4) −40, −14, −12, 23, 47
5) −57, −54, 32, 36, 59
6) −34, −19, 26, 47, 68
7) 36, 18, −10, −16, −18
8) 94, 34, 27, −12, −24
9) 50, 42, −2, −13, −21
10) 86, 46, 37, −16, −20
11) 88, 75, −18, −26, −59
12) 14, 3, −25, −30, −65

Common Core Subject Test Mathematics Grade 6

Integers and Absolute Value

1) 2	8) 2	15) 32	22) −2
2) 27	9) 54	16) 17	23) 32
3) 20	10) 4	17) 42	24) 19
4) 14	11) 11	18) 46	25) 28
5) 6	12) 88	19) 1	26) 10
6) 55	13) 0	20) 40	27) 28
7) 16	14) 79	21) 2	28) 88

Factoring Numbers

1) 1, 2, 3, 4, 6, 12
2) 1, 2, 4, 8, 16
3) 1, 2, 4, 7, 14, 28
4) 1, 2, 17, 34
5) 1, 5, 19, 95
6) 1, 2, 4, 7, 8, 14, 28, 56
7) 1, 5, 13, 65
8) 1, 2, 5, 7, 10, 14, 35, 70
9) 1, 5, 25
10) 1, 2, 3, 4, 6, 8, 12, 16, 24, 48
11) 1, 3, 9, 27
12) 1, 3, 7, 9, 21, 63
13) 1, 2, 3, 4, 6, 8, 9, 12, 18, 24, 36, 72
14) 1, 3, 5, 15
15) 1, 2, 4, 5, 8, 10, 16, 20, 40, 80
16) 2×5
17) 2×13
18) $2 \times 2 \times 5$
19) $2 \times 3 \times 5$
20) $2 \times 2 \times 2 \times 5$
21) $2 \times 2 \times 11$
22) 5×11
23) $2 \times 3 \times 13$
24) $2 \times 2 \times 2 \times 2 \times 2 \times 3$

Prime Factorization

1) 2. 3	9) 2. 29	17) 3. 23
2) 7. 7	10) 2. 31	18) 2. 2. 19
3) 2. 2. 3. 5	11) 3. 5. 5	19) 2. 43
4) 2. 2	12) 2. 2. 2. 11	20) 2. 2. 23
5) 2. 23	13) 3. 31	21) 3. 3. 11
6) 3. 19	14) 2. 2. 5. 5	22) 7. 11
7) 2. 3. 3. 3	15) 2. 2. 17	23) 2. 3. 3. 5
8) 2. 19	16) 2. 3. 3. 5	24) 2. 37

Common Core Subject Test Mathematics Grade 6

Divisibility Rules

1) 8   <u>2</u> 3 <u>4</u> 5 6 7 <u>8</u> 9 10
2) 18   <u>2</u> <u>3</u> 4 5 <u>6</u> 7 8 <u>9</u> 10
3) 55   2 3 4 <u>5</u> 6 7 8 9 10
4) 45   2 <u>3</u> 4 <u>5</u> 6 7 8 <u>9</u> 10
5) 20   <u>2</u> 3 <u>4</u> <u>5</u> 6 7 8 9 <u>10</u>
6) 9   2 <u>3</u> 4 5 6 7 8 <u>9</u> 10
7) 21   2 <u>3</u> 4 5 6 <u>7</u> 8 9 10
8) 28   <u>2</u> 3 <u>4</u> 5 6 <u>7</u> 8 9 10
9) 36   <u>2</u> <u>3</u> <u>4</u> 5 <u>6</u> 7 8 <u>9</u> 10
10) 40   <u>2</u> 3 <u>4</u> <u>5</u> 6 7 <u>8</u> 9 <u>10</u>
11) 39   2 <u>3</u> 4 5 6 7 8 9 10
12) 51   2 <u>3</u> 4 5 6 7 8 9 10

Greatest Common Factor

1) 2
2) 1
3) 3
4) 1
5) 5
6) 8
7) 6
8) 3
9) 6
10) 4
11) 2
12) 4
13) 5
14) 2
15) 9
16) 2
17) 14
18) 8
19) 1
20) 5
21) 1
22) 16
23) 6
24) 9

Least Common Multiple

1) 18
2) 45
3) 80
4) 36
5) 54
6) 42
7) 30
8) 56
9) 21
10) 40
11) 75
12) 63
13) 44
14) 56
15) 56
16) 200
17) 156
18) 22
19) 180
20) 105
21) 162
22) 270
23) 90
24) 75

Common Core Subject Test Mathematics Grade 6

Common Core Subject Test Mathematics Grade 6

Chapter 3 :
Fractions

Topics that you will practice in this chapter:

- ✓ Simplifying Fractions
- ✓ Adding and Subtracting Fractions
- ✓ Multiplying and Dividing Fractions
- ✓ Adding and Subtract Mixed Numbers
- ✓ Multiplying and Dividing Mixed Numbers

"A Man is like a fraction whose numerator is what he is and whose denominator is what he thinks of himself. The larger the denominator, the smaller the fraction." –Tolstoy

Common Core Subject Test Mathematics Grade 6

Simplifying Fractions

✎ **Simplify each fraction to its lowest terms.**

1) $\frac{5}{10} =$

2) $\frac{28}{35} =$

3) $\frac{27}{36} =$

4) $\frac{40}{80} =$

5) $\frac{14}{56} =$

6) $\frac{32}{48} =$

7) $\frac{52}{65} =$

8) $\frac{15}{60} =$

9) $\frac{80}{160} =$

10) $\frac{55}{77} =$

11) $\frac{28}{112} =$

12) $\frac{32}{64} =$

13) $\frac{63}{72} =$

14) $\frac{81}{90} =$

15) $\frac{35}{105} =$

16) $\frac{25}{70} =$

17) $\frac{80}{280} =$

18) $\frac{12}{81} =$

19) $\frac{36}{186} =$

20) $\frac{240}{540} =$

21) $\frac{70}{560} =$

✎ **Find the answer for each problem.**

22) Which of the following fractions equal to $\frac{3}{4}$? ____

   A. $\frac{60}{90}$ B. $\frac{43}{104}$ C. $\frac{48}{64}$ D. $\frac{150}{300}$

23) Which of the following fractions equal to $\frac{5}{8}$? ____

   A. $\frac{125}{200}$ B. $\frac{115}{200}$ C. $\frac{50}{100}$ D. $\frac{30}{90}$

24) Which of the following fractions equal to $\frac{3}{7}$? ____

   A. $\frac{58}{116}$ B. $\frac{54}{126}$ C. $\frac{270}{167}$ D. $\frac{42}{63}$

WWW.MathNotion.Com

Common Core Subject Test Mathematics Grade 6

Adding and Subtracting Fractions

✏️ **Find the sum.**

1) $\dfrac{5}{9} + \dfrac{4}{9} =$

2) $\dfrac{1}{2} + \dfrac{1}{7} =$

3) $\dfrac{3}{8} + \dfrac{1}{4} =$

4) $\dfrac{3}{5} + \dfrac{1}{2} =$

5) $\dfrac{1}{4} + \dfrac{3}{5} =$

6) $\dfrac{7}{8} + \dfrac{3}{8} =$

7) $\dfrac{1}{2} + \dfrac{7}{10} =$

8) $\dfrac{2}{5} + \dfrac{2}{3} =$

9) $\dfrac{5}{7} + \dfrac{2}{3} =$

10) $\dfrac{7}{12} + \dfrac{3}{4} =$

11) $\dfrac{5}{6} + \dfrac{2}{5} =$

12) $\dfrac{1}{12} + \dfrac{2}{3} =$

✏️ **Find the difference.**

13) $\dfrac{1}{3} - \dfrac{1}{6} =$

14) $\dfrac{3}{4} - \dfrac{1}{8} =$

15) $\dfrac{1}{2} - \dfrac{1}{3} =$

16) $\dfrac{1}{4} - \dfrac{1}{5} =$

17) $\dfrac{5}{8} - \dfrac{2}{3} =$

18) $\dfrac{1}{4} - \dfrac{1}{7} =$

19) $\dfrac{5}{6} - \dfrac{1}{9} =$

20) $\dfrac{3}{4} - \dfrac{1}{6} =$

21) $\dfrac{7}{8} - \dfrac{1}{12} =$

22) $\dfrac{8}{15} - \dfrac{3}{5} =$

23) $\dfrac{3}{12} - \dfrac{1}{14} =$

24) $\dfrac{10}{13} - \dfrac{7}{26} =$

25) $\dfrac{6}{7} - \dfrac{3}{4} =$

26) $\dfrac{4}{5} - \dfrac{1}{8} =$

27) $\dfrac{4}{7} - \dfrac{2}{35} =$

28) $\dfrac{9}{16} - \dfrac{2}{8} =$

29) $\dfrac{8}{9} - \dfrac{7}{18} =$

30) $\dfrac{1}{2} - \dfrac{4}{9} =$

Common Core Subject Test Mathematics Grade 6

Multiplying and Dividing Fractions

✎ Find the value of each expression in lowest terms.

1) $\dfrac{1}{5} \times \dfrac{15}{5} =$

2) $\dfrac{9}{12} \times \dfrac{4}{9} =$

3) $\dfrac{1}{16} \times \dfrac{8}{10} =$

4) $\dfrac{1}{24} \times \dfrac{8}{10} =$

5) $\dfrac{1}{5} \times \dfrac{1}{4} =$

6) $\dfrac{7}{9} \times \dfrac{1}{7} =$

7) $\dfrac{6}{7} \times \dfrac{1}{3} =$

8) $\dfrac{2}{8} \times \dfrac{2}{8} =$

9) $\dfrac{5}{8} \times \dfrac{3}{5} =$

10) $\dfrac{4}{7} \times \dfrac{1}{8} =$

11) $\dfrac{7}{15} \times \dfrac{5}{7} =$

12) $\dfrac{3}{10} \times \dfrac{5}{9} =$

✎ Find the value of each expression in lowest terms.

13) $\dfrac{1}{4} \div \dfrac{1}{8} =$

14) $\dfrac{1}{10} \div \dfrac{1}{5} =$

15) $\dfrac{3}{4} \div \dfrac{1}{5} =$

16) $\dfrac{1}{3} \div \dfrac{5}{6} =$

17) $\dfrac{1}{7} \div \dfrac{8}{42} =$

18) $\dfrac{3}{4} \div \dfrac{1}{6} =$

19) $\dfrac{2}{7} \div \dfrac{7}{13} =$

20) $\dfrac{1}{24} \div \dfrac{3}{16} =$

21) $\dfrac{7}{12} \div \dfrac{5}{6} =$

22) $\dfrac{22}{18} \div \dfrac{11}{9} =$

23) $\dfrac{9}{35} \div \dfrac{3}{7} =$

24) $\dfrac{2}{7} \div \dfrac{8}{21} =$

25) $\dfrac{1}{9} \div \dfrac{2}{5} =$

26) $\dfrac{5}{12} \div \dfrac{3}{5} =$

27) $\dfrac{3}{20} \div \dfrac{1}{6} =$

28) $\dfrac{8}{20} \div \dfrac{3}{4} =$

29) $\dfrac{5}{6} \div \dfrac{2}{9} =$

30) $\dfrac{5}{11} \div \dfrac{3}{4} =$

WWW.MathNotion.Com

Common Core Subject Test Mathematics Grade 6

Adding and Subtracting Mixed Numbers

✏️ **Find the sum.**

1) $3\frac{1}{3} + 2\frac{1}{6} =$

2) $4\frac{1}{2} + 3\frac{1}{2} =$

3) $3\frac{3}{8} + 1\frac{1}{8} =$

4) $2\frac{1}{4} + 2\frac{1}{3} =$

5) $3\frac{5}{6} + 2\frac{7}{12} =$

6) $5\frac{4}{15} + 3\frac{3}{5} =$

7) $2\frac{1}{3} + 4\frac{3}{7} =$

8) $3\frac{1}{2} + 4\frac{2}{5} =$

9) $5\frac{2}{5} + 6\frac{3}{7} =$

10) $8\frac{5}{16} + 6\frac{1}{12} =$

✏️ **Find the difference.**

11) $3\frac{1}{4} - 1\frac{3}{4} =$

12) $6\frac{3}{5} - 4\frac{2}{5} =$

13) $4\frac{1}{3} - 3\frac{1}{9} =$

14) $7\frac{1}{7} - 5\frac{1}{2} =$

15) $5\frac{1}{3} - 2\frac{1}{12} =$

16) $8\frac{1}{5} - 4\frac{1}{3} =$

17) $9\frac{1}{4} - 6\frac{1}{8} =$

18) $11\frac{7}{15} - 8\frac{3}{5} =$

19) $14\frac{5}{6} - 11\frac{3}{5} =$

20) $18\frac{2}{7} - 14\frac{1}{5} =$

21) $9\frac{1}{3} - 4\frac{1}{4} =$

22) $6\frac{1}{8} - 4\frac{1}{16} =$

23) $19\frac{3}{8} - 15\frac{1}{3} =$

24) $11\frac{1}{9} - 8\frac{1}{8} =$

25) $17\frac{1}{7} - 11\frac{1}{5} =$

26) $16\frac{2}{9} - 9\frac{5}{7} =$

WWW.MathNotion.Com

Common Core Subject Test Mathematics Grade 6

Multiplying and Dividing Mixed Numbers

✎ **Find the product.**

1) $5\frac{1}{2} \times 2\frac{1}{4} =$

2) $5\frac{1}{3} \times 4\frac{1}{3} =$

3) $5\frac{3}{4} \times 6\frac{1}{4} =$

4) $3\frac{1}{3} \times 2\frac{3}{5} =$

5) $4\frac{8}{10} \times 1\frac{1}{24} =$

6) $6\frac{2}{7} \times 1\frac{1}{11} =$

7) $8\frac{2}{3} \times 3\frac{1}{2} =$

8) $3\frac{4}{7} \times 2\frac{1}{5} =$

9) $5\frac{2}{8} \times 4\frac{1}{6} =$

10) $7\frac{3}{3} \times 1\frac{3}{8} =$

✎ **Find the quotient.**

11) $2\frac{2}{5} \div 4\frac{1}{5} =$

12) $4\frac{1}{6} \div 3\frac{1}{3} =$

13) $6\frac{1}{3} \div 1\frac{1}{2} =$

14) $7\frac{1}{10} \div 2\frac{2}{5} =$

15) $3\frac{1}{3} \div 1\frac{1}{9} =$

16) $1\frac{1}{10} \div 4\frac{1}{2} =$

17) $1\frac{3}{16} \div 5\frac{1}{4} =$

18) $4\frac{1}{3} \div 4\frac{3}{4} =$

19) $9\frac{1}{3} \div 2\frac{1}{4} =$

20) $15\frac{1}{3} \div 5\frac{1}{2} =$

21) $4\frac{1}{6} \div 1\frac{1}{5} =$

22) $1\frac{1}{18} \div 1\frac{2}{9} =$

23) $4\frac{2}{7} \div 1\frac{3}{10} =$

24) $7\frac{1}{3} \div 2\frac{2}{11} =$

25) $8\frac{2}{5} \div 1\frac{1}{6} =$

26) $9\frac{1}{3} \div 2\frac{1}{7} =$

WWW.MathNotion.Com

Common Core Subject Test Mathematics Grade 6

Answers of Worksheets

Simplifying Fractions

1) $\frac{1}{2}$
2) $\frac{4}{5}$
3) $\frac{3}{4}$
4) $\frac{1}{2}$
5) $\frac{1}{4}$
6) $\frac{2}{3}$
7) $\frac{4}{5}$
8) $\frac{1}{4}$
9) $\frac{1}{2}$
10) $\frac{5}{7}$
11) $\frac{1}{4}$
12) $\frac{1}{2}$
13) $\frac{7}{8}$
14) $\frac{9}{10}$
15) $\frac{1}{3}$
16) $\frac{5}{14}$
17) $\frac{2}{7}$
18) $\frac{4}{27}$
19) $\frac{6}{31}$
20) $\frac{4}{9}$
21) $\frac{1}{8}$
22) C
23) A
24) B

Adding and Subtracting Fractions

1) $\frac{9}{9} = 1$
2) $\frac{9}{14}$
3) $\frac{5}{8}$
4) $1\frac{1}{10}$
5) $\frac{17}{20}$
6) $1\frac{1}{4}$
7) $1\frac{1}{5}$
8) $1\frac{1}{15}$
9) $1\frac{8}{21}$
10) $1\frac{1}{3}$
11) $1\frac{7}{30}$
12) $\frac{3}{4}$
13) $\frac{1}{6}$
14) $\frac{5}{8}$
15) $\frac{1}{6}$
16) $\frac{1}{20}$
17) $-\frac{1}{24}$
18) $\frac{3}{28}$
19) $\frac{13}{18}$
20) $\frac{7}{12}$
21) $\frac{19}{24}$
22) $-\frac{1}{15}$
23) $\frac{5}{28}$
24) $\frac{1}{2}$
25) $\frac{3}{28}$
26) $\frac{27}{40}$
27) $\frac{18}{35}$
28) $\frac{5}{16}$
29) $\frac{1}{2}$
30) $\frac{1}{18}$

Multiplying and Dividing Fractions

1) $\frac{3}{5}$
2) $\frac{1}{3}$
3) $\frac{1}{20}$
4) $\frac{1}{30}$
5) $\frac{1}{20}$
6) $\frac{1}{9}$
7) $\frac{2}{7}$
8) $\frac{1}{16}$
9) $\frac{3}{8}$
10) $\frac{1}{14}$
11) $\frac{1}{3}$
12) $\frac{1}{6}$
13) 2
14) $\frac{1}{2}$
15) $3\frac{3}{4}$
16) $\frac{2}{5}$

WWW.MathNotion.Com

Common Core Subject Test Mathematics Grade 6

17) $\frac{3}{4}$

18) $4\frac{1}{2}$

19) $\frac{26}{49}$

20) $\frac{2}{9}$

21) $\frac{7}{10}$

22) 1

23) $\frac{3}{5}$

24) $\frac{3}{4}$

25) $\frac{5}{18}$

26) $\frac{25}{36}$

27) $\frac{9}{10}$

28) $\frac{8}{15}$

29) $3\frac{3}{4}$

30) $\frac{20}{33}$

Adding and Subtracting Mixed Numbers

1) $5\frac{1}{2}$

2) 8

3) $4\frac{1}{2}$

4) $4\frac{7}{12}$

5) $6\frac{5}{12}$

6) $8\frac{13}{15}$

7) $6\frac{16}{21}$

8) $7\frac{9}{10}$

9) $11\frac{29}{35}$

10) $14\frac{19}{48}$

11) $1\frac{1}{2}$

12) $2\frac{1}{5}$

13) $1\frac{2}{9}$

14) $1\frac{9}{14}$

15) $3\frac{1}{4}$

16) $3\frac{13}{15}$

17) $3\frac{1}{8}$

18) $2\frac{13}{15}$

19) $3\frac{7}{30}$

20) $4\frac{3}{35}$

21) $5\frac{1}{12}$

22) $2\frac{1}{16}$

23) $4\frac{1}{24}$

24) $2\frac{71}{72}$

25) $5\frac{33}{35}$

26) $6\frac{32}{63}$

Multiplying and Dividing Mixed Numbers

1) $12\frac{3}{8}$

2) $23\frac{1}{9}$

3) $35\frac{15}{16}$

4) $8\frac{2}{3}$

5) 5

6) $6\frac{6}{7}$

7) $30\frac{1}{3}$

8) $7\frac{6}{7}$

9) $21\frac{7}{8}$

10) 11

11) $\frac{4}{7}$

12) $1\frac{1}{4}$

13) $4\frac{2}{9}$

14) $2\frac{23}{24}$

15) 3

16) $\frac{11}{45}$

17) $\frac{19}{84}$

18) $\frac{52}{57}$

19) $4\frac{4}{27}$

20) $2\frac{26}{33}$

21) $3\frac{17}{36}$

22) $\frac{19}{22}$

23) $3\frac{27}{91}$

24) $3\frac{13}{36}$

25) $7\frac{1}{5}$

26) $4\frac{16}{45}$

Common Core Subject Test Mathematics Grade 6

Chapter 4 :
Decimals

Topics that you will practice in this chapter:

- ✓ Adding and Subtracting Decimals
- ✓ Multiplying and Dividing Decimals
- ✓ Comparing Decimals
- ✓ Rounding Decimals

"The study of mathematics, like the Nile, begins in minuteness but ends in magnificence." – Charles Caleb Colton

Common Core Subject Test Mathematics Grade 6

Adding and Subtracting Decimals

✎ Add and subtract decimals.

1) 35.19 − 24.28 = _____

2) 34.29 + 42.58 = _____

3) 61.20 + 33.75 = _____

4) 38.72 − 21.68 = _____

5) 57.39 + 26.54 = _____

6) 70.24 − 42.35 = _____

7) 86.09 − 35.14 = _____

8) 54.51 + 32.66 = _____

9) 114.21 − 88.69 = _____

✎ Find the missing number.

10) ___ + 2.8 = 5.4

11) 4.1 + ___ = 5.88

12) 6.45 + ___ = 8

13) 7.25 − ___ = 3.40

14) ___ − 2.35 = 4.25

15) ___ − 19.85 = 6.54

16) 22.15 + ___ = 28.95

17) ___ − 37.16 = 9.42

18) ___ + 24.50 = 34.19

19) 72.40 + ___ = 125.20

Common Core Subject Test Mathematics Grade 6

Multiplying and Dividing Decimals

✎ **Find the product.**

1) $0.5 \times 0.6 =$

2) $3.3 \times 0.4 =$

3) $1.28 \times 0.5 =$

4) $0.35 \times 0.6 =$

5) $1.85 \times 0.6 =$

6) $0.24 \times 0.5 =$

7) $5.25 \times 1.4 =$

8) $18.5 \times 4.6 =$

9) $15.4 \times 6.8 =$

10) $19.5 \times 2.6 =$

11) $32.2 \times 1.5 =$

12) $78.4 \times 4.5 =$

✎ **Find the quotient.**

13) $1.85 \div 10 =$

14) $74.6 \div 100 =$

15) $3.6 \div 3 =$

16) $9.6 \div 0.4 =$

17) $15.5 \div 0.5 =$

18) $32.8 \div 0.2 =$

19) $22.15 \div 1{,}000 =$

20) $53.55 \div 0.7 =$

21) $322.2 \div 0.2 =$

22) $50.67 \div 0.18 =$

23) $77.4 \div 0.8 =$

24) $27.93 \div 0.03 =$

WWW.MathNotion.Com

Comparing Decimals

✎ **Write the correct comparison symbol (>, < or =).**

1) 0.70 ☐ 0.070

2) 0.049 ☐ 0.49

3) 5.090 ☐ 5.09

4) 2.57 ☐ 2.05

5) 9.03 ☐ 0.930

6) 6.06 ☐ 6.6

7) 7.02 ☐ 7.020

8) 3.04 ☐ 3.2

9) 3.61 ☐ 3.245

10) 0.986 ☐ 0.0986

11) 17.24 ☐ 17.240

12) 0.759 ☐ 0.81

13) 9.040 ☐ 9.40

14) 5.73 ☐ 5.213

15) 9.44 ☐ 9.404

16) 7.17 ☐ 7.170

17) 4.85 ☐ 4.085

18) 9.041 ☐ 9.40

19) 3.033 ☐ 3.030

20) 4.97 ☐ 4.970

Common Core Subject Test Mathematics Grade 6

Rounding Decimals

✎ **Round each decimal to the nearest whole number.**

1) 28.12   3) 16.22   5) 7.95

2) 6.9   4) 8.5   6) 52.7

✎ **Round each decimal to the nearest tenth.**

7) 31.761   9) 94.729   11) 13.219

8) 14.421   10) 77.89   12) 59.89

✎ **Round each decimal to the nearest hundredth.**

13) 8.428   15) 55.3786 17) 62.241

14) 23.812   16) 231.912 18) 19.447

✎ **Round each decimal to the nearest thousandth.**

19) 15.54324   21) 243.8652   23) 67.1983

20) 34.62586   22) 80.4529   24) 72.36788

Common Core Subject Test Mathematics Grade 6

Convert Fraction to Decimal

✏ Write each as a decimal.

1) $\dfrac{50}{100} =$

2) $\dfrac{46}{100} =$

3) $\dfrac{8}{50} =$

4) $\dfrac{8}{32} =$

5) $\dfrac{8}{72} =$

6) $\dfrac{56}{100} =$

7) $\dfrac{4}{50} =$

8) $\dfrac{31}{48} =$

9) $\dfrac{27}{300} =$

10) $\dfrac{15}{55} =$

11) $\dfrac{16}{32} =$

12) $\dfrac{6}{16} =$

13) $\dfrac{3}{10} =$

14) $\dfrac{18}{250} =$

15) $\dfrac{24}{80} =$

16) $\dfrac{30}{40} =$

17) $\dfrac{68}{100} =$

18) $\dfrac{7}{35} =$

19) $\dfrac{87}{100} =$

20) $\dfrac{1}{100} =$

21) $\dfrac{6}{36} =$

22) $\dfrac{2}{80} =$

WWW.MathNotion.Com

Common Core Subject Test Mathematics Grade 6

Convert Decimal to Percent

✏️ **Write each as a percent.**

1) 0.187 =

2) 0.19 =

3) 2.6 =

4) 0.017 =

5) 0.009 =

6) 0.786 =

7) 0.245 =

8) 0.57 =

9) 0.002 =

10) 0.205 =

11) 0.324 =

12) 84.9 =

13) 3.015 =

14) 0.7 =

15) 2.35 =

16) 0.0367 =

17) 0.0043 =

18) 0.960 =

19) 6.68 =

20) 0.484 =

21) 8.957 =

22) 0.879 =

23) 2.7 =

24) 0.9 =

25) 3.6 =

26) 26.8 =

27) 1.01 =

28) 0.006 =

Common Core Subject Test Mathematics Grade 6

Convert Fraction to Percent

✏ Write each as a percent.

1) $\dfrac{1}{4} =$

2) $\dfrac{3}{8} =$

3) $\dfrac{7}{14} =$

4) $\dfrac{15}{35} =$

5) $\dfrac{12}{28} =$

6) $\dfrac{17}{68} =$

7) $\dfrac{8}{11} =$

8) $\dfrac{14}{30} =$

9) $\dfrac{6}{50} =$

10) $\dfrac{12}{48} =$

11) $\dfrac{5}{34} =$

12) $\dfrac{27}{10} =$

13) $\dfrac{24}{80} =$

14) $\dfrac{16}{25} =$

15) $\dfrac{16}{58} =$

16) $\dfrac{2}{22} =$

17) $\dfrac{32}{88} =$

18) $\dfrac{21}{36} =$

19) $\dfrac{18}{92} =$

20) $\dfrac{6}{60} =$

21) $\dfrac{24}{600} =$

22) $\dfrac{720}{360} =$

WWW.MathNotion.Com

Common Core Subject Test Mathematics Grade 6

Answers of Worksheets

Adding and Subtracting Decimals

1) 10.91	6) 27.89	11) 1.78	16) 6.8
2) 76.87	7) 50.95	12) 1.55	17) 46.58
3) 94.95	8) 87.17	13) 3.85	18) 9.69
4) 17.04	9) 25.52	14) 6.6	19) 52.8
5) 83.93	10) 2.6	15) 26.39	

Multiplying and Dividing Decimals

1) 0.3	7) 7.35	13) 0.185	19) 0.02215
2) 1.32	8) 85.1	14) 0.746	20) 76.5
3) 0.64	9) 104.72	15) 1.2	21) 1,611
4) 0.21	10) 50.7	16) 24	22) 281.5
5) 1.11	11) 48.3	17) 31	23) 96.75
6) 0.12	12) 352.8	18) 164	24) 931

Comparing Decimals

1) >	6) <	11) =	16) =
2) <	7) =	12) <	17) >
3) =	8) <	13) <	18) <
4) >	9) >	14) >	19) >
5) >	10) >	15) >	20) =

Rounding Decimals

1) 28	9) 94.7	17) 62.24	
2) 7	10) 77.9	18) 19.45	
3) 16	11) 13.2	19) 15.543	
4) 9	12) 59.9	20) 34.626	
5) 8	13) 8.43	21) 243.865	
6) 53	14) 23.81	22) 80.453	
7) 31.8	15) 55.38	23) 67.198	
8) 14.4	16) 231.91	24) 72.368	

Convert Fraction to Decimal

1) 0.5	2) 0.46	3) 0.16

Common Core Subject Test Mathematics Grade 6

4) 0.25
5) 0.11
6) 0.56
7) 0.08
8) 0.646
9) 0.09
10) 0.27

11) 0.5
12) 0.375
13) 0.3
14) 0.072
15) 0.3
16) 0.75
17) 0.68

18) 0.2
19) 0.87
20) 0.01
21) 0.166
22) 0.025

Convert Decimal to Percent

1) 18.7%
2) 19%
3) 260%
4) 1.7%
5) 0.9%
6) 78.6%
7) 24.5%
8) 57%
9) 0.2%
10) 20.5%

11) 32.4%
12) 8,490%
13) 301.5%
14) 70%
15) 235%
16) 3.67%
17) 0.43%
18) 96%
19) 668%
20) 48.4%

21) 895.7%
22) 87.9%
23) 270%
24) 90%
25) 360%
26) 2,680%
27) 101%
28) 0.6%

Convert Fraction to Percent

1) 25%
2) 37.5%
3) 50%
4) 42.86%
5) 29.31%
6) 25%
7) 72.72%
8) 46.66%

9) 12%
10) 25%
11) 14.7%
12) 2.7%
13) 30%
14) 64%
15) 27.58%
16) 9.09%

17) 36.36%
18) 58.33%
19) 19.56%
20) 10%
21) 4%
22) 200%

Common Core Subject Test Mathematics Grade 6

Chapter 5:
Proportions, Ratios, and Percent

Topics that you will practice in this chapter:

- ✓ Simplifying Ratios
- ✓ Proportional Ratios
- ✓ Similarity and Ratios
- ✓ Ratio and Rates Word Problems
- ✓ Percentage Calculations
- ✓ Percent Problems
- ✓ Discount, Tax and Tip

Without mathematics, there's nothing you can do. Everything around you is mathematics. Everything around you is numbers." – Shakuntala Devi

Common Core Subject Test Mathematics Grade 6

Simplifying Ratios

✍ **Reduce each ratio.**

1) 15: 20 = ___ : ___
2) 7: 70 = ___ : ___
3) 16: 28 = ___ : ___
4) 7: 21 = ___ : ___
5) 4: 40 = ___ : ___
6) 6: 48 = ___ : ___
7) 16: 64 = ___ : ___
8) 10: 25 = ___ : ___

9) 8: 48 = ___ : ___
10) 49: 63 = ___ : ___
11) 18: 27 = ___ : ___
12) 35: 10 = ___ : ___
13) 90: 9 = ___ : ___
14) 24: 32 = ___ : ___
15) 7: 56 = ___ : ___
16) 45: 63 = ___ : ___

17) 56: 72 = ___ : ___
18) 26: 13 = ___ : ___
19) 15: 45 = ___ : ___
20) 28: 4 = ___ : ___
21) 24: 48 = ___ : ___
22) 30: 24 = ___ : ___
23) 70: 140 = ___ : ___
24) 6: 180 = ___ : ___

✍ **Write each ratio as a fraction in simplest form.**

25) 6: 12 =
26) 30: 50 =
27) 15: 35 =
28) 9: 27 =
29) 8: 24 =
30) 18: 84 =
31) 7: 14 =

32) 7: 35 =
33) 40: 96 =
34) 12: 54 =
35) 44: 52 =
36) 12: 27 =
37) 15: 180 =
38) 39: 143 =

39) 20: 300 =
40) 30: 120 =
41) 56: 42 =
42) 26: 130 =
43) 66: 123 =
44) 70: 630 =
45) 75: 125 =

WWW.MathNotion.Com

Common Core Subject Test Mathematics Grade 6

Proportional Ratios

✎ Fill in the blanks; Calculate each proportion.

1) $3:8 = \_\_ : 48$

2) $2:5 = 20:\_\_$

3) $1:9 = \_\_ : 81$

4) $6:7 = 12:\_\_$

5) $9:2 = 63:\_\_$

6) $8:7 = \_\_ : 49$

7) $20:3 = \_\_ : 15$

8) $1:3 = \_\_ : 75$

9) $7:6 = \_\_ : 60$

10) $8:5 = \_\_ : 45$

11) $3:10 = 60:\_\_$

12) $6:11 = 42:\_\_$

✎ State if each pair of ratios form a proportion.

13) $\frac{3}{20}$ and $\frac{9}{60}$

14) $\frac{1}{7}$ and $\frac{6}{42}$

15) $\frac{3}{7}$ and $\frac{24}{56}$

16) $\frac{4}{9}$ and $\frac{12}{18}$

17) $\frac{1}{9}$ and $\frac{12}{81}$

18) $\frac{7}{8}$ and $\frac{21}{28}$

19) $\frac{9}{13}$ and $\frac{27}{39}$

20) $\frac{1}{8}$ and $\frac{8}{64}$

21) $\frac{6}{19}$ and $\frac{30}{85}$

22) $\frac{5}{9}$ and $\frac{40}{81}$

23) $\frac{9}{14}$ and $\frac{108}{168}$

24) $\frac{15}{23}$ and $\frac{360}{552}$

✎ Calculate each proportion.

25) $\frac{20}{25} = \frac{32}{x}$, $x = $ ___

26) $\frac{1}{8} = \frac{32}{x}$, $x = $ ___

27) $\frac{15}{5} = \frac{21}{x}$, $x = $ ___

28) $\frac{1}{7} = \frac{x}{294}$, $x = $ ___

29) $\frac{7}{9} = \frac{x}{81}$, $x = $ ___

30) $\frac{1}{5} = \frac{13}{x}$, $x = $ ___

31) $\frac{9}{5} = \frac{36}{x}$, $x = $ ___

32) $\frac{6}{13} = \frac{48}{x}$, $x = $ ___

33) $\frac{5}{8} = \frac{x}{88}$, $x = $ ___

34) $\frac{4}{15} = \frac{x}{240}$, $x = $ ___

35) $\frac{9}{19} = \frac{x}{266}$, $x = $ ___

36) $\frac{7}{15} = \frac{x}{270}$, $x = $ ___

WWW.MathNotion.Com

Similarity and Ratios

✎ **Each pair of figures is similar. Find the missing side.**

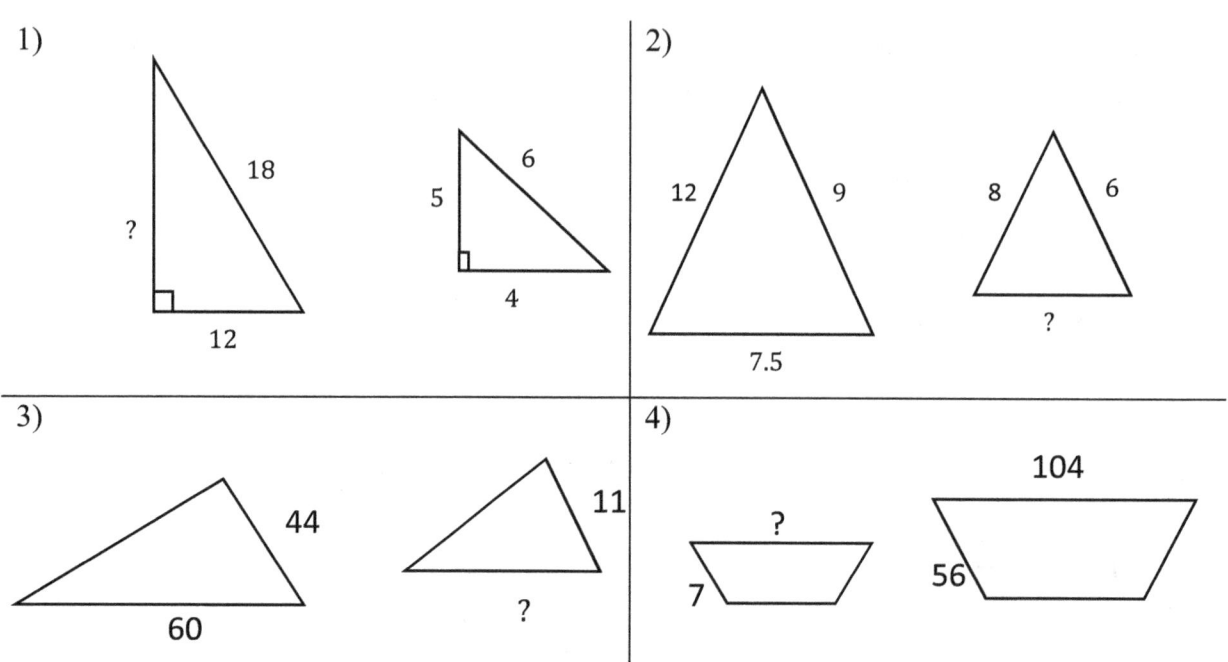

✎ **Calculate.**

5) Two rectangles are similar. The first is 24 feet wide and 120 feet long. The second is 30 feet wide. What is the length of the second rectangle? _____

6) Two rectangles are similar. One is 5 meters by 36 meters. The longer side of the second rectangle is 90 meters. What is the other side of the second rectangle? _____

7) A building casts a shadow 25 ft long. At the same time a girl 10 ft tall casts a shadow 5 ft long. How tall is the building? _____

8) The scale of a map of Texas is 4 inches: 32 miles. If you measure the distance from Dallas to Martin County as 38.4 inches, approximately how far is Martin County from Dallas? _____

Common Core Subject Test Mathematics Grade 6

Ratio and Rates Word Problems

✍ **Find the answer for each word problem.**

1) Mason has 24 red cards and 36 green cards. What is the ratio of Mason's red cards to his green cards? _____

2) In a party, 45 soft drinks are required for every 54 guests. If there are 378 guests, how many soft drinks is required? _____

3) In Mason's class, 42 of the students are tall and 24 are short. In Michael's class 84 students are tall and 48 students are short. Which class has a higher ratio of tall to short students? _____

4) The price of 5 apples at the Quick Market is $4.6. The price of 7 of the same apples at Walmart is $5.95. Which place is the better buy? _____

5) The bakers at a Bakery can make 90 bagels in 3 hours. How many bagels can they bake in 24 hours? What is that rate per hour? _____

6) You can buy 5 cans of green beans at a supermarket for $5.75. How much does it cost to buy 45 cans of green beans? _____

7) The ratio of boys to girls in a class is 4: 7. If there are 32 boys in the class, how many girls are in that class? _____

8) The ratio of red marbles to blue marbles in a bag is 3: 7. If there are 50 marbles in the bag, how many of the marbles are red? _____

WWW.MathNotion.Com

Common Core Subject Test Mathematics Grade 6

Percentage Calculations

✏ Calculate the given percent of each value.

1) 3% of 60 = ____
2) 20% of 32 = ____
3) 4% of 72 = ____
4) 16% of 32 = ____
5) 25% of 124 = ____
6) 35% of 56 = ____

7) 15% of 20 = ____
8) 14% of 150 = ____
9) 80% of 50 = ____
10) 12% of 115 = ____
11) 72% of 250 = ____
12) 52% of 500 = ____

13) 70% of 400 = ____
14) 27% of 145 = ____
15) 90% of 64 = ____
16) 60% of 55 = ____
17) 22% of 210 = ____
18) 8% of 235 = ____

✏ Calculate the percent of each given value.

19) ____% of 25 = 5
20) ____% of 40 = 20
21) ____% of 25 = 2
22) ____% of 50 = 16
23) ____% of 250 = 5

24) ____% of 40 = 32
25) ____% of 125 = 20
26) ____% of 700 = 49
27) ____% of 350 = 49
28) ____% of 500 = 210

✏ Calculate each percent problem.

29) A Cinema has 250 seats. 60 seats were sold for the current movie. What percent of seats are empty? ____ %

30) There are 68 boys and 92 girls in a class. 75% of the students in the class take the bus to school. How many students do not take the bus to school? ____

WWW.MathNotion.Com

Common Core Subject Test Mathematics Grade 6

Percent Problems

✎ **Calculate each problem.**

1) 9 is what percent of 45? ____%

2) 60 is what percent of 120? ____%

3) 10 is what percent of 200? ____%

4) 15 is what percent of 125? ____%

5) 10 is what percent of 400? ____%

6) 66 is what percent of 55? ____%

7) 40 is what percent of 160? ____%

8) 40 is what percent of 50? ____%

9) 120 is what percent of 800? ____%

10) 78 is what percent of 120? ____%

11) 36 is what percent of 144? ____%

12) 17 is what percent of 85? ____%

13) 90 is what percent of 900? ____%

14) 36 is what percent of 16? ____%

15) 63 is what percent of 14? ____%

16) 18 is what percent of 60? ____%

17) 126 is what percent of 200? ____%

18) 232 is what percent of 40? ____%

✎ **Calculate each percent word problem.**

19) There are 40 employees in a company. On a certain day, 25 were present. What percent showed up for work? ____%

20) A metal bar weighs 60 ounces. 25% of the bar is gold. How many ounces of gold are in the bar? _____

21) A crew is made up of 12 women; the rest are men. If 15% of the crew are women, how many people are in the crew? _____

22) There are 40 students in a class and 8 of them are girls. What percent are boys? ____%

23) The Royals softball team played 400 games and won 280 of them. What percent of the games did they lose? ____%

Common Core Subject Test Mathematics Grade 6

Discount, Tax and Tip

🖎 **Find the selling price of each item.**

1) Original price of a computer: $420
Tax: 8% Selling price: $_____

2) Original price of a laptop: $280
Tax: 4% Selling price: $_____

3) Original price of a sofa: $820
Tax: 5% Selling price: $_____

4) Original price of a car: $15,800
Tax: 3.6%   Selling price: $_____

5) Original price of a Table: $250
Tax: 9% Selling price: $_____

6) Original price of a house: $630,000
Tax: 1.8%   Selling price: _____

7) Original price of a tablet: $450
Discount: 30% Selling price: $___

8) Original price of a chair: $390
Discount: 8%   Selling price: $___

9) Original price of a book: $75
Discount: 42% Selling price: $__

10) Original price of a cellphone: $820
Discount: 23% Selling price: $__

11) Food bill: $45
Tip: 15%   Price: $_____

12) Food bill: $32
Tipp: 20%   Price: $_____

13) Food bill: $90
Tip: 35%   Price: $_____

14) Food bill: $42
Tipp: 12%   Price: $_____

🖎 **Find the answer for each word problem.**

15) Nicolas hired a moving company. The company charged $500 for its services, and Nicolas gives the movers a 40% tip. How much does Nicolas tip the movers? $_____

16) Mason has lunch at a restaurant and the cost of his meal is $90. Mason wants to leave a 25% tip. What is Mason's total bill including tip? $_____

17) The sales tax in Texas is 19.80% and an item costs $350. How much is the tax? $_____

18) The price of a table at Best Buy is $680. If the sales tax is 5%, what is the final price of the table including tax? $_____

WWW.MathNotion.Com

Common Core Subject Test Mathematics Grade 6

Answers of Worksheets

Simplifying Ratios

1) 3:4
2) 1:10
3) 4:7
4) 1:3
5) 1:10
6) 1:8
7) 2:8
8) 2:5
9) 1:6
10) 7:9
11) 2:3
12) 7:2
13) 10:1
14) 3:4
15) 1:8
16) 5:7
17) 7:9
18) 2:1
19) 1:3
20) 7:1
21) 1:2
22) 5:4
23) 1:2
24) 1:30
25) $\frac{1}{2}$
26) $\frac{3}{5}$
27) $\frac{3}{7}$
28) $\frac{1}{3}$
29) $\frac{1}{3}$
30) $\frac{3}{14}$
31) $\frac{1}{2}$
32) $\frac{1}{5}$
33) $\frac{5}{12}$
34) $\frac{2}{9}$
35) $\frac{11}{13}$
36) $\frac{4}{9}$
37) $\frac{1}{12}$
38) $\frac{3}{11}$
39) $\frac{1}{15}$
40) $\frac{1}{4}$
41) $\frac{4}{3}$
42) $\frac{1}{5}$
43) $\frac{22}{41}$
44) $\frac{1}{9}$
45) $\frac{3}{5}$

Proportional Ratios

1) 18
2) 50
3) 9
4) 14
5) 14
6) 56
7) 100
8) 25
9) 70
10) 72
11) 200
12) 77
13) Yes
14) Yes
15) Yes
16) No
17) No
18) No
19) Yes
20) Yes
21) No
22) No
23) Yes
24) Yes
25) 40
26) 256
27) 7
28) 42
29) 63
30) 65
31) 20
32) 104
33) 55
34) 64
35) 126
36) 126

Similarity and ratios

1) 15
2) 5
3) 15
4) 13
5) 150 feet
6) 12.5 meters
7) 50 feet
8) 307.2 miles

Ratio and Rates Word Problems

1) 2:3
2) 315

WWW.MathNotion.Com

Common Core Subject Test Mathematics Grade 6

3) The ratio for both classes is 7 to 4.
4) Walmart is a better buy.
5) 720, the rate is 30 per hour.
6) $51.75
7) 56
8) 15

Percentage Calculations

1) 1.8
2) 6.4
3) 2.88
4) 5.12
5) 31
6) 19.6
7) 3
8) 21
9) 40
10) 13.8
11) 180
12) 260
13) 280
14) 39.15
15) 57.6
16) 33
17) 46.2
18) 18.8
19) 20%
20) 50%
21) 8%
22) 32%
23) 2%
24) 80%
25) 16%
26) 7%
27) 14%
28) 42%
29) 76%
30) 40

Percent Problems

1) 20%
2) 50%
3) 5%
4) 12%
5) 2.5%
6) 120%
7) 25%
8) 80%
9) 15%
10) 65%
11) 25%
12) 20%
13) 10%
14) 225%
15) 450%
16) 30%
17) 63%
18) 580%
19) 62.5%
20) 15 ounces
21) 80
22) 80%
23) 30%

Discount, Tax and Tip

1) $453.60
2) $291.20
3) $861.00
4) $16,368.80
5) $272.50
6) $641,340
7) $315.00
8) $358.80
9) $43.50
10) $631.40
11) $51.75
12) $38.40
13) $121.50
14) $47.04
15) $200.00
16) $112.50
17) $69.30
18) $714.00

Common Core Subject Test Mathematics Grade 6

Chapter 6:
Exponents and Radicals Expressions

Topics that you will practice in this chapter:

- ✓ Adding and Subtracting Exponents
- ✓ Multiplication Property of Exponents
- ✓ Zero and Negative Exponents
- ✓ Division Property of Exponents
- ✓ Powers of Products and Quotients
- ✓ Negative Exponents and Negative Bases
- ✓ Scientific Notation
- ✓ Square Roots

Mathematics is no more computation than typing is literature.

– John Allen Paulos

Common Core Subject Test Mathematics Grade 6

Adding and Subtracting Exponents

✎ Solve each problem.

1) $3^2 + 2^5 =$

2) $x^6 + x^6 =$

3) $3b^2 - 2b^2 =$

4) $3 + 4^3 =$

5) $8 - 4^2 =$

6) $4 + 7^1 =$

7) $2x^3 + 3x^3 =$

8) $10^2 + 3^5 =$

9) $4^5 - 2^4 =$

10) $5^2 - 6^0 =$

11) $1^2 - 3^0 =$

12) $7^1 + 2^3 =$

13) $6^1 - 5^3 =$

14) $3^3 + 3^3 =$

15) $9^2 - 8^2 =$

16) $0^{73} + 0^{54} =$

17) $2^2 - 3^2 =$

18) $7^3 - 7^1 =$

19) $8^2 - 6^2 =$

20) $4^2 + 3^2 =$

21) $2^3 + 4^3 =$

22) $10 + 3^3 =$

23) $6x^5 + 8x^5 =$

24) $8^0 + 4^2 =$

25) $3^2 + 3^2 =$

26) $10^2 + 5^2 =$

27) $(\frac{1}{2})^2 + (\frac{1}{2})^2 =$

28) $9^2 + 3^2 =$

WWW.MathNotion.Com

Multiplication Property of Exponents

✏ Simplify and write the answer in exponential form.

1) $4 \times 4^5 =$

2) $8^4 \times 8 =$

3) $7^3 \times 7^3 =$

4) $9^2 \times 9^2 =$

5) $2^2 \times 2^4 \times 2 =$

6) $5 \times 5^3 \times 5^3 =$

7) $4^3 \times 4^2 \times 4 \times 4 =$

8) $5x \times x =$

9) $x^3 \times x^3 =$

10) $x^7 \times x^2 =$

11) $x^4 \times x^3 \times x^2 =$

12) $10x \times 3x =$

13) $4x^3 \times 4x^3 =$

14) $7x^3 \times x =$

15) $3x^2 \times 4x^2 \times x^2 =$

16) $5x^4 \times x^4 =$

17) $2x^8 \times 2x =$

18) $6x \times x^5 =$

19) $4x^2 \times 6x^6 =$

20) $5yx^3 \times 4x =$

21) $7x^3 \times y^5 x^7 =$

22) $y^2 x^3 \times y^5 x^4 =$

23) $3x^5 \times 4x^3 y^4 =$

24) $4x^4 \times 9x^2 y^5 =$

25) $5x^3 y^4 \times 6x^8 y^2 =$

26) $8x^3 y^6 \times 4xy^3 =$

27) $2xy^5 \times 6x^3 y^3 =$

28) $4x^5 y^2 \times 4x^2 y^8 =$

29) $7x \times 3y^8 x^2 \times y^5 =$

30) $x^3 \times 2y^3 x^4 \times 2y =$

31) $3yx^4 \times 3y^4 x \times 3xy^3 =$

32) $6y^3 \times 2y^2 x^4 \times 10yx^5 =$

Common Core Subject Test Mathematics Grade 6

Zero and Negative Exponents

✎ **Evaluate the following expressions.**

1) $1^{-5} =$

2) $4^{-1} =$

3) $0^{10} =$

4) $1^{15} =$

5) $5^{-2} =$

6) $3^{-3} =$

7) $9^{-1} =$

8) $10^{-2} =$

9) $12^{-2} =$

10) $2^{-5} =$

11) $3^{-4} =$

12) $2^{-4} =$

13) $6^{-3} =$

14) $10^{-3} =$

15) $30^{-1} =$

16) $15^{-2} =$

17) $4^{-3} =$

18) $2^{-7} =$

19) $5^{-3} =$

20) $4^{-4} =$

21) $3^{-5} =$

22) $10^{-4} =$

23) $2^{-10} =$

24) $8^{-3} =$

25) $20^{-2} =$

26) $14^{-2} =$

27) $9^{-3} =$

28) $100^{-2} =$

29) $5^{-4} =$

30) $4^{-6} =$

31) $\left(\frac{1}{4}\right)^{-3} =$

32) $\left(\frac{1}{6}\right)^{-2} =$

33) $\left(\frac{1}{7}\right)^{-2} =$

34) $\left(\frac{2}{3}\right)^{-3} =$

35) $\left(\frac{1}{13}\right)^{-2} =$

36) $\left(\frac{7}{12}\right)^{-2} =$

37) $\left(\frac{1}{6}\right)^{-3} =$

38) $\left(\frac{1}{300}\right)^{-2} =$

39) $\left(\frac{2}{9}\right)^{-2} =$

40) $\left(\frac{7}{5}\right)^{-1} =$

41) $\left(\frac{13}{23}\right)^{0} =$

42) $\left(\frac{1}{4}\right)^{-5} =$

WWW.MathNotion.Com

Division Property of Exponents

✎ Simplify.

1) $\dfrac{5^6}{5^7} =$

2) $\dfrac{8^8}{8^6} =$

3) $\dfrac{4^5}{4} =$

4) $\dfrac{3}{3^5} =$

5) $\dfrac{x}{x^6} =$

6) $\dfrac{3 \times 3^2}{3^2 \times 3^5} =$

7) $\dfrac{9^4}{9^2} =$

8) $\dfrac{10 \times 10^9}{10^2 \times 10^7} =$

9) $\dfrac{7^5 \times 7^7}{7^4 \times 7^8} =$

10) $\dfrac{15x}{30x^6} =$

11) $\dfrac{3x^9}{4x^4} =$

12) $\dfrac{15x^8}{10x^9} =$

13) $\dfrac{42x^5}{6y^9} =$

14) $\dfrac{36y^8}{4x^4y^5} =$

15) $\dfrac{2x^7}{9x} =$

16) $\dfrac{49x^8y^6}{7x^9} =$

17) $\dfrac{48x^2}{24x^6y^{12}} =$

18) $\dfrac{30yx^5}{6yx^7} =$

19) $\dfrac{19x^7y}{38x^{12}y^4} =$

20) $\dfrac{9x^8}{63x^8} =$

21) $\dfrac{9x^{-9}}{4x^{-3}} =$

Common Core Subject Test Mathematics Grade 6

Powers of Products and Quotients

✏️ **Simplify.**

1) $(4^3)^2 =$

2) $(2^3)^4 =$

3) $(2 \times 2^3)^2 =$

4) $(5 \times 5^5)^6 =$

5) $(19^4 \times 19^2)^3 =$

6) $(2^3 \times 2^4)^4 =$

7) $(5 \times 5^2)^2 =$

8) $(4^4)^4 =$

9) $(8x^5)^2 =$

10) $(3x^2y^4)^4 =$

11) $(7x^5y^2)^2 =$

12) $(5x^4y^4)^3 =$

13) $(2x^3y^3)^5 =$

14) $(10x^3y^4)^3 =$

15) $(13y^3y)^2 =$

16) $(5x^6x^4)^2 =$

17) $(6x^7y^6)^3 =$

18) $(12x^5x^7)^2 =$

19) $(2x^4 \times 2x)^4 =$

20) $(2x^4y^3)^5 =$

21) $(15x^7y^2)^2 =$

22) $(8x^3y^5)^3 =$

23) $(3x \times 2y^2)^4 =$

24) $\left(\dfrac{4x}{x^5}\right)^2 =$

25) $\left(\dfrac{x^4y^5}{x^3y^5}\right)^9 =$

26) $\left(\dfrac{36xy}{6x^5}\right)^3 =$

27) $\left(\dfrac{x^7}{x^8y^2}\right)^6 =$

28) $\left(\dfrac{xy^4}{x^3y^6}\right)^{-3} =$

29) $\left(\dfrac{5xy^8}{x^3}\right)^2 =$

30) $\left(\dfrac{xy^6}{2xy^3}\right)^{-4} =$

WWW.MathNotion.Com

Common Core Subject Test Mathematics Grade 6

Negative Exponents and Negative Bases

✏️ **Simplify.**

1) $-9^{-1} =$

2) $-9^{-2} =$

3) $-2^{-5} =$

4) $-x^{-7} =$

5) $11x^{-1} =$

6) $-8x^{-3} =$

7) $-12x^{-5} =$

8) $-9x^{-8}y^{-6} =$

9) $32x^{-5}y^{-1} =$

10) $10a^{-9}b^{-3} =$

11) $-17x^4y^{-6} =$

12) $-\dfrac{25}{x^{-5}} =$

13) $-\dfrac{13x}{a^{-7}} =$

14) $\left(-\dfrac{1}{3}\right)^{-4} =$

15) $\left(-\dfrac{3}{4}\right)^{-2} =$

16) $-\dfrac{14}{a^{-6}b^{-3}} =$

17) $-\dfrac{7x}{x^{-8}} =$

18) $-\dfrac{a^{-9}}{b^{-5}} =$

19) $-\dfrac{11}{x^{-5}} =$

20) $\dfrac{8b}{-16c^{-6}} =$

21) $\dfrac{12ab}{a^{-4}b^{-3}} =$

22) $-\dfrac{8n^{-4}}{32p^{-7}} =$

23) $\dfrac{16ab^{-6}}{-6c^{-5}} =$

24) $\left(\dfrac{10a}{5c}\right)^{-4} =$

25) $\left(-\dfrac{12x}{4yz}\right)^{-3} =$

26) $\dfrac{8ab^{-7}}{-5c^{-3}} =$

27) $\left(-\dfrac{x^4}{x^5}\right)^{-5} =$

28) $\left(-\dfrac{x^{-2}}{7x^3}\right)^{-2} =$

29) $\left(-\dfrac{x^{-4}}{x^2}\right)^{-6} =$

Scientific Notation

✎ Write each number in scientific notation.

1) $0.223 =$ 11) $8,000,000 =$

2) $0.09 =$ 12) $0.00009 =$

3) $4.5 =$ 13) $2,158,000 =$

4) $900 =$ 14) $0.0039 =$

5) $2,000 =$ 15) $0.000075 =$

6) $0.006 =$ 16) $4,300,000 =$

7) $33 =$ 17) $130,000 =$

8) $9,400 =$ 18) $4,000,000,000 =$

9) $1,470 =$ 19) $0.00009 =$

10) $52,000 =$ 20) $0.0039 =$

✎ Write each number in standard notation.

21) $4 \times 10^{-1} =$ 26) $5.5 \times 10^5 =$

22) $1.2 \times 10^{-3} =$ 27) $3.2 \times 10^4 =$

23) $2.7 \times 10^5 =$ 28) $3.88 \times 10^6 =$

24) $6 \times 10^{-4} =$ 29) $7 \times 10^{-6} =$

25) $3.6 \times 10^{-3} =$ 30) $4.2 \times 10^{-7} =$

Common Core Subject Test Mathematics Grade 6

Square Roots

✏ **Find the value each square root.**

1) $\sqrt{16} =$ ___

2) $\sqrt{25} =$ ___

3) $\sqrt{1} =$ ___

4) $\sqrt{64} =$ ___

5) $\sqrt{0} =$ ___

6) $\sqrt{196} =$ ___

7) $\sqrt{4} =$ ___

8) $\sqrt{256} =$ ___

9) $\sqrt{36} =$ ___

10) $\sqrt{289} =$ ___

11) $\sqrt{169} =$ ___

12) $\sqrt{144} =$ ___

13) $\sqrt{100} =$ ___

14) $\sqrt{1,600} =$ ___

15) $\sqrt{2,500} =$ ___

16) $\sqrt{324} =$ ___

17) $\sqrt{529} =$ ___

18) $\sqrt{20} =$ ___

19) $\sqrt{625} =$ ___

20) $\sqrt{18} =$ ___

21) $\sqrt{50} =$ ___

22) $\sqrt{1,024} =$ ___

23) $\sqrt{160} =$ ___

24) $\sqrt{32} =$ ___

✏ **Evaluate.**

25) $\sqrt{4} \times \sqrt{25} =$ _____

26) $\sqrt{36} \times \sqrt{49} =$ _____

27) $\sqrt{6} \times \sqrt{6} =$ _____

28) $\sqrt{13} \times \sqrt{13} =$ _____

29) $2\sqrt{5} \times 3\sqrt{5} =$ _____

30) $\sqrt{12} \times \sqrt{3} =$ _____

31) $\sqrt{13} + \sqrt{13} =$ _____

32) $\sqrt{10} + 2\sqrt{10} =$ _____

33) $12\sqrt{7} - 10\sqrt{7} =$ _____

34) $4\sqrt{10} \times 2\sqrt{10} =$ _____

35) $5\sqrt{3} \times 8\sqrt{3} =$ _____

36) $6\sqrt{3} - \sqrt{12} =$ _____

WWW.MathNotion.Com

Common Core Subject Test Mathematics Grade 6

Answers of Worksheets

Add and Subtract Exponents.

1) 41
2) $2x^6$
3) b^2
4) 67
5) -8
6) 11
7) $5x^3$
8) 343
9) 1,008
10) 24
11) 0
12) 15
13) -119
14) 54
15) 17
16) 0
17) -5
18) 336
19) 28
20) 25
21) 72
22) 37
23) $14x^5$
24) 17
25) 18
26) 125
27) $\frac{1}{2}$
28) 90

Multiplication Property of Exponents

1) 4^6
2) 8^5
3) 7^6
4) 9^4
5) 2^7
6) 5^7
7) 4^7
8) $5x^2$
9) x^6
10) x^9
11) x^9
12) $30x^2$
13) $16x^6$
14) $7x^4$
15) $12x^6$
16) $5x^8$
17) $4x^9$
18) $6x^6$
19) $24x^8$
20) $20x^4y$
21) $7x^{10}y^5$
22) x^7y^7
23) $12x^8y^4$
24) $36x^6y^5$
25) $30x^{11}y^6$
26) $32x^4y^9$
27) $12x^4y^8$
28) $16x^7y^{10}$
29) $21x^3y^{13}$
30) $4x^7y^4$
31) $27x^6y^8$
32) $120x^9y^6$

Zero and Negative Exponents

1) 1
2) $\frac{1}{4}$
3) 0
4) 1
5) $\frac{1}{25}$
6) $\frac{1}{27}$
7) $\frac{1}{9}$
8) $\frac{1}{100}$
9) $\frac{1}{144}$
10) $\frac{1}{32}$
11) $\frac{1}{81}$
12) $\frac{1}{16}$
13) $\frac{1}{216}$
14) $\frac{1}{1,000}$
15) $\frac{1}{30}$
16) $\frac{1}{225}$
17) $\frac{1}{64}$
18) $\frac{1}{128}$
19) $\frac{1}{125}$
20) $\frac{1}{256}$
21) $\frac{1}{243}$
22) $\frac{1}{10,000}$
23) $\frac{1}{1,024}$
24) $\frac{1}{512}$
25) $\frac{1}{400}$

WWW.MathNotion.Com

Common Core Subject Test Mathematics Grade 6

26) $\frac{1}{196}$ 30) $\frac{1}{4,096}$ 35) 169   40) $\frac{5}{7}$

27) $\frac{1}{729}$ 31) 64   36) $\frac{144}{49}$ 41) 1

28) $\frac{1}{10,000}$ 32) 36   37) 216   42) 1,024

29) $\frac{1}{625}$ 33) 49   38) 90,000

34) $\frac{27}{8}$ 39) $\frac{81}{4}$

Division Property of Exponents

1) $\frac{1}{5}$ 7) 9^2 13) $\frac{7x^5}{y^9}$ 18) $\frac{5}{x^2}$

2) 8^2 8) 10   14) $\frac{9y^3}{x^4}$ 19) $\frac{1}{2x^5y^3}$

3) 4^4 9) 1   15) $\frac{2x^6}{9}$ 20) $\frac{1}{7}$

4) $\frac{1}{3^4}$ 10) $\frac{1}{2x^5}$ 16) $\frac{7y^6}{x}$ 21) $\frac{9}{4x^6}$

5) $\frac{1}{x^5}$ 11) $\frac{3x^5}{4}$ 17) $\frac{2}{x^4 y^{12}}$

6) $\frac{1}{3^4}$ 12) $\frac{3}{2x}$

Powers of Products and Quotients

1) 4^6 12) $125x^{12}y^{12}$ 23) $1,296x^4y^8$

2) 2^{12} 13) $32x^{15}y^{15}$ 24) $\frac{16}{x^8}$

3) 2^8 14) $1,000x^9y^{12}$ 25) x^9

4) 5^{36} 15) $169y^8$ 26) $\frac{216y^3}{x^{12}}$

5) 19^{18} 16) $25x^{20}$

6) 2^{28} 17) $216x^{21}y^{18}$ 27) $\frac{1}{x^6y^{12}}$

7) 5^6 18) $144x^{24}$ 28) x^6y^6

8) 4^{16} 19) $256x^{20}$ 29) $\frac{25y^{16}}{x^4}$

9) $64x^{10}$ 20) $32x^{20}y^{15}$

10) $81x^8y^{16}$ 21) $225x^{14}y^4$ 30) $\frac{16}{y^{12}}$

11) $49x^{10}y^4$ 22) $512x^9y^{15}$

Negative Exponents and Negative Bases

1) $-\frac{1}{9}$ 4) $-\frac{1}{x^7}$ 7) $-\frac{12}{x^5}$

2) $-\frac{1}{81}$ 5) $\frac{11}{x}$ 8) $-\frac{9}{x^8y^6}$

3) $-\frac{1}{32}$ 6) $-\frac{8}{x^3}$ 9) $\frac{32}{x^5y}$

WWW.MathNotion.Com

Common Core Subject Test Mathematics Grade 6

10) $\frac{10}{a^9 b^3}$

11) $-\frac{17x^4}{y^6}$

12) $-25x^5$

13) $-13xa^7$

14) 81

15) $\frac{16}{9}$

16) $-14a^6 b^3$

17) $-7x^9$

18) $-\frac{b^5}{a^9}$

19) $-11x^5$

20) $-\frac{bc^6}{2}$

21) $12a^5 b^4$

22) $-\frac{p^7}{4n^4}$

23) $-\frac{8ac^5}{3b^6}$

24) $\frac{c^4}{16a^4}$

25) $\frac{y^3 z^3}{27x^3}$

26) $-\frac{8ac^3}{5b^7}$

27) $-x^5$

28) $49x^{10}$

29) x^{36}

Scientific Notation

1) 2.23×10^{-1}
2) 9×10^{-2}
3) 4.5×10^0
4) 9×10^2
5) 2×10^3
6) 6×10^{-3}
7) 3.3×10^1
8) 9.4×10^3
9) 1.47×10^3
10) 5.2×10^4
11) 8×10^6
12) 9×10^{-5}
13) 2.158×10^6
14) 3.9×10^{-3}
15) 7.5×10^{-5}
16) 4.3×10^6
17) 1.3×10^5
18) 4×10^9
19) 9×10^{-5}
20) 3.9×10^{-3}
21) 0.4
22) 0.0012
23) 270,000
24) 0.0006
25) 0.0036
26) 550,000
27) 32,000
28) 3,880,000
29) 0.000007
30) 0.00000042

Square Roots

1) 4
2) 5
3) 1
4) 8
5) 0
6) 14
7) 2
8) 16
9) 6
10) 17
11) 13
12) 12
13) 10
14) 40
15) 50
16) 18
17) 23
18) $2\sqrt{5}$
19) 25
20) $3\sqrt{2}$
21) $5\sqrt{2}$
22) 32
23) $4\sqrt{10}$
24) $4\sqrt{2}$
25) 10
26) 42
27) 6
28) 13
29) 30
30) 6
31) $2\sqrt{13}$
32) $3\sqrt{10}$
33) $2\sqrt{7}$
34) 80
35) 120
36) $4\sqrt{3}$

WWW.MathNotion.Com

Common Core Subject Test Mathematics Grade 6

Chapter 7 : Measurements

Topics that you will learn in this chapter:

- ✓ Reference Measurement
- ✓ Metric Length
- ✓ Customary Length
- ✓ Metric Capacity
- ✓ Customary Capacity
- ✓ Metric Weight and Mass
- ✓ Customary Weight and Mass
- ✓ Temperature
- ✓ Time

"It's not that I'm so smart, it's just that I stay with problems longer." -Albert Einstein

Common Core Subject Test Mathematics Grade 6

Reference Measurement

LENGTH	
Customary	**Metric**
1 mile (mi) = 1,760 yards (yd)	1 kilometer (km) = 1,000 meters (m)
1 yard (yd) = 3 feet (ft)	1 meter (m) = 100 centimeters (cm)
1 foot (ft) = 12 inches (in.)	1 centimeter(cm) = 10 millimeters(mm)

VOLUME AND CAPACITY	
Customary	**Metric**
1 gallon (gal) = 4 quarts (qt)	1 liter (L) = 1,000 milliliters (mL)
1 quart (qt) = 2 pints (pt.)	
1 pint (pt.) = 2 cups (c)	
1 cup (c) = 8 fluid ounces (Fl oz)	

WEIGHT AND MASS	
Customary	**Metric**
1 ton (T) = 2,000 pounds (lb.)	1 kilogram (kg) = 1,000 grams (g)
1 pound (lb.) = 16 ounces (oz)	1 gram (g) = 1,000 milligrams (mg)

Time
1 year = 12 months
1 year = 52 weeks
1 week = 7 days
1 day = 24 hours
1 hour = 60 minutes
1 minute = 60 seconds

Common Core Subject Test Mathematics Grade 6

Metric Length Measurement

✏️ **Convert to the units.**

1) 3×10^3 mm = _____ cm
2) 0.95 m = _____ mm
3) 0.08 m = _____ cm
4) 2.25 km = _____ m
5) 7,800 mm = _____ m
6) 9,100 cm = _____ m
7) 5.83 m = _____ cm

8) 2×10^5 mm = _____ cm
9) 8×10^3 mm = _____ m
10) 0.003 km = _____ mm
11) 0.7 km = _____ m
12) 0.011 m = _____ cm
13) 125×10^5 m = _____ km
14) 78×10^4 m = _____ km

Customary Length Measurement

✏️ **Convert to the units.**

1) 15 ft = _____ in
2) 1.5 ft = _____ in
3) 4.8 yd = _____ ft
4) 0.82 yd = _____ ft
5) 17×10^{-3} yd = _____ in
6) 0.5 mi = _____ in
7) 1,746 in = _____ yd
8) 3.24 in = _____ yd

9) 3,960 yd = _____ mi
10) 42.55 yd = _____ in
11) 5×10^{-2} mi = _____ yd
12) 87,120 ft = _____ mi
13) 2.52 in = _____ ft
14) 29.3 yd = _____ feet
15) 0.612 in = _____ ft
16) 1.3 mi = _____ ft

WWW.MathNotion.Com

Common Core Subject Test Mathematics Grade 6

Metric Capacity Measurement

✍ **Convert the following measurements.**

1) 1.58 l = _____ ml

2) 0.504 l = _____ ml

3) 3.04 l = _____ ml

4) 0.005 l = _____ ml

5) 121.56 l = _____ ml

6) 0.0459 l = _____ ml

7) 4.2×10^5 ml = _____ l

8) 3.12×10^3 ml = _____ l

9) $1,889 \times 10^2$ ml = _____ l

10) 250 ml = _____ l

11) 656,160 ml = _____ l

12) 0.54×10^4 ml = _____ l

Customary Capacity Measurement

✍ **Convert the following measurements.**

1) 0.7 gal = _____ qt.

2) 3.2 gal = _____ pt.

3) 0.75 gal = _____ c.

4) 15.5 pt. = _____ c

5) 18.2 c = _____ fl oz

6) 9.02 qt = _____ pt.

7) 1.05 qt = _____ c

8) 158 pt. = _____ c

9) 9.6×10^3 c = _____ gal

10) 203.2 pt. = _____ gal

11) 12.4 qt = _____ gal

12) 115.6 pt. = _____ qt

13) 4,880 c = _____ qt

14) 113.6 c = _____ pt.

15) 0.036 qt = _____ gal

16) 522.4 pt. = _____ qt

17) 5.8 gal = _____ pt.

18) 0.002 qt = _____ c

19) 672 c = _____ gal

20) 72.96 fl oz = _____ c

WWW.MathNotion.Com

Common Core Subject Test Mathematics Grade 6

Metric Weight and Mass Measurement

✎ Convert.

1) 0.712 kg = _____ g

2) 54.01 kg = _____ g

3) 9.8×10^{-5} kg = _____ g

4) 0.012 kg = _____ g

5) 120.02 kg = _____ g

6) 1.199 kg = _____ g

7) 0.0055 kg = _____ g

8) 9×10^4 g = _____ kg

9) 3.5×10^5 g = _____ kg

10) 0.008×10^4 g = _____ kg

11) 15,010 g = _____ kg

12) 12.1×10^4 g = _____ kg

13) 4,155,200 g = _____ kg

14) 402×10^2 g = _____ kg

Customary Weight and Mass Measurement

✎ Convert.

1) 36×10^2 lb. = _____ T

2) 0.022×10^4 lb. = _____ T

3) 215,000 lb. = _____ T

4) 12,600 lb. = _____ T

5) 0.015 lb. = _____ oz

6) 1.6 lb. = _____ oz

7) 0.021 lb. = _____ oz

8) 5.2 T = _____ lb.

9) 6.8×10^{-3} T = _____ lb.

10) 156×10^{-2} T = _____ lb.

11) 0.017 T = _____ lb.

12) 1.085 T = _____ oz

13) 0.006 T = _____ oz

14) 209.92 oz = _____ lb.

WWW.MathNotion.Com

Temperature

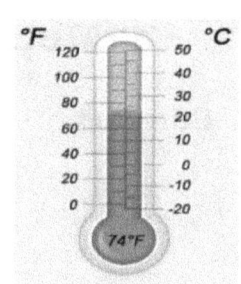

✎ **Convert Fahrenheit into Celsius.**

1) 35.6°F = ___ °C

2) 54.5°F = ___ °C

3) −25.6°F = ___ °C

4) 62.6°F = ___ °C

5) 120.2°F = ___ °C

6) 174.2°F = ___ °C

7) 51.8°F = ___ °C

8) 193.1°F = ___ °C

9) 221°F = ___ °C

10) 60.44°F = ___ °C

11) 48.2°F = ___ °C

12) 134.6°F = ___ °C

✎ **Convert Celsius into Fahrenheit.**

13) 18.2°C = ___ °F

14) 88.8°C = ___ °F

15) 250°C = ___ °F

16) 52°C = ___ °F

17) 10°C = ___ °F

18) −24°C = ___ °F

19) 6°C = ___ °F

20) 15.6°C = ___ °F

21) 33°C = ___ °F

22) 61°C = ___ °F

23) 113.5°C = ___ °F

24) 28°C = ___ °F

Common Core Subject Test Mathematics Grade 6

Time

✎ **Convert to the units.**

1) 18.5 hr. = _____ min

2) 26 year = _____ week

3) 0.2 hr. = _____ sec

4) 12.5 min = _____ sec

5) 1.2×10^5 min = _____ hr

6) 1,460 day = _____ year

7) 3 year = _____ hr.

8) 51 day = _____ hr

9) 5 day = _____ min

10) 552 min = _____ hr

11) 32.25 year = _____ month

12) 6,480 sec = _____ min

13) 264 hr = _____ day

14) 22 weeks = _____ day

✎ **How much time has passed?**

1) From 1:45 A.M. to 4:55 A.M.: ____ hours and ____ minutes.

2) From 1:25 A.M. to 6:05 A.M.: ____ hours and ____ minutes.

3) It's 7:15 P.M. What time was 4 hours ago? _____ O'clock

4) 3:05 A.M to 6:55 AM: ____ hours and ____ minutes.

5) 3:45 A.M to 5:15 AM: ____ hours and ____ minutes.

6) 8:05 A.M. to 11:20 AM. = ____ hour(s) and ____ minutes.

7) 10:55 A.M. to 1:25 PM. = ____ hour(s) and ____ minutes

8) 6:18 A.M. to 6:52 A.M. = ____ minutes

9) 3:54 A.M. to 4:08 A.M. = ____ minutes

WWW.MathNotion.Com

Common Core Subject Test Mathematics Grade 6

Answers of Worksheets

Metric length

1) 300 cm
2) 950 mm
3) 8 cm
4) 2,250 m
5) 7.8 m
6) 91 m
7) 583 cm
8) 20,000 cm
9) 8 m
10) 3,000 mm
11) 700 m
12) 1.1 cm
13) 12,500 km
14) 780km

Customary Length

1) 180
2) 18
3) 14.4
4) 2.46
5) 0.612
6) 31,680
7) 48.5
8) 0.09
9) 2.25
10) 1,531.8
11) 88
12) 16.5
13) 0.21
14) 87.9
15) 0.051
16) 6,864

Metric Capacity

1) 1,580 ml
2) 504 ml
3) 3,040 ml
4) 5 ml
5) 121,560 ml
6) 45.9 ml
7) 420 L
8) 3.12 L
9) 188.9 L
10) 0.25L
11) 656.16 L
12) 5.4 L

Customary Capacity

1) 2.8 qt
2) 25.6 pt.
3) 12 c
4) 31 c
5) 145.6 fl oz
6) 18.04 pt.
7) 4.2 c
8) 316 c
9) 600 gal
10) 25.4 gal
11) 3.1 gal
12) 57.8 qt
13) 1,220qt
14) 56.8 pt.
15) 0.009 gal
16) 261.2 qt
17) 46.4 pt.
18) 0.008 c
19) 42 gal
20) 9.12 c

Metric Weight and Mass

1) 712 g
2) 54,010 g
3) 0.098 g
4) 12 g
5) 120,020 g
6) 1,199 g
7) 5.5 g
8) 90 kg
9) 350 kg
10) 0.08 kg
11) 15.01 kg
12) 121 kg
13) 4,155.2 kg
14) 40.2 kg

WWW.MathNotion.Com

Common Core Subject Test Mathematics Grade 6

Customary Weight and Mass

1) 1.8 T	6) 25.6 oz	11) 34 lb.
2) 0.11 T	7) 0.336 oz	12) 34,720 oz
3) 107.5 T	8) 10,400 lb.	13) 192 oz
4) 6.3 T	9) 13.6 lb.	14) 13.12 lb
5) 0.24 oz	10) 3,120 lb.	

Temperature

1) 2°C	9) 105°C	17) 50°F
2) 12.5°C	10) 15.8°C	18) −11.2°F
3) −32°C	11) 9°C	19) 42.8°F
4) 17°C	12) 57°C	20) 60.08°F
5) 49°C	13) 64.76°F	21) 91.4°F
6) 79°C	14) 191.84°F	22) 141.8°F
7) 11°C	15) 482°F	23) 236.3°F
8) 89.5°C	16) 125.6°F	24) 82.4°F

Time - Convert.

1) 1,110 min	6) 4 year	11) 387 months
2) 1,352 weeks	7) 26,280 hr	12) 108 min
3) 720 sec	8) 1,224 hr	13) 11 days
4) 750 sec	9) 7,200 min	14) 154 days
5) 2,000 hr	10) 9.2 hr	

Time - Gap

1) 3:10	4) 3:50	7) 2:30
2) 4:40	5) 1:30	8) 34 minutes
3) 3:15 P.M.	6) 3:15	9) 14 minutes

Common Core Subject Test Mathematics Grade 6

Chapter 8 :
Algebraic Expressions

Topics that you will practice in this chapter:

- ✓ Find a rule!
- ✓ Translate Phrases into an Algebraic Statement
- ✓ Simplifying Variable Expressions
- ✓ The Distributive Property
- ✓ Evaluating One Variable Expressions
- ✓ Evaluating Two Variables Expressions
- ✓ Combining like Terms

Mathematics is, as it were, a sensuous logic, and relates to philosophy as do the arts, music, and plastic art to poetry. — *K. Shegel*

Common Core Subject Test Mathematics Grade 6

Find a Rule!

✎ **Complete the output.**

1- **Rule:** the output is $x - 10.5$

Input	x	15	18	27	32.25	48.5
Output	y					

1) **Rule:** the output is $x \times 5\frac{1}{3}$

Input	x	3	9	15	21	33
Output	y					

2- **Rule:** the output is $x \div 9$

Input	x	513	387	342	198	126
Output	y					

✎ **Find a rule to write an expression.**

3- **Rule:** _____

Input	x	4	14	19	24
Output	y	10	35	47.5	60

4- **Rule:** _____

Input	x	5	13	19.6	34.5
Output	y	14.4	22.4	29	43.9

5- **Rule:** _____

Input	x	72	96	132	230.4
Output	y	9	12	16.5	28.8

Common Core Subject Test Mathematics Grade 6

Translate Phrases into an Algebraic Statement

✏️ Write an algebraic expression for each phrase.

1) 9 multiplied by x. _____

2) Subtract 11 from y. _____

3) 19 divided by x. _____

4) 38 decreased by y. _____

5) Add y to 40. _____

6) The square of 6. _____

7) x raised to the fifth power. _____

8) The sum of six and a number. _____

9) The difference between fifty-seven and y. _____

10) The quotient of nine and a number. _____

11) The quotient of the square of x and 25. _____

12) The difference between x and 6 is 19. _____

13) 10 times a reduced by the square of b. _____

14) Subtract the product of a and b from 41. _____

WWW.MathNotion.Com

Common Core Subject Test Mathematics Grade 6

Simplifying Variable Expressions

✍ **Simplify each expression.**

1) $3(x+5) =$

2) $(-4)(7x-5) =$

3) $11x + 5 - 6x =$

4) $-4 - 2x^2 - 6x^2 =$

5) $7 + 13x^2 + 3 =$

6) $3x^2 + 7x + 15x^2 =$

7) $3x^2 - 12x^2 + 4x =$

8) $4x^2 - 8x - 2x =$

9) $6x + 7(3 - 4x) =$

10) $8x + 4(15x - 3) =$

11) $6(-3x - 9) - 17 =$

12) $-11x^2 - (-5x) =$

13) $2x + 7 + 5 - 8x =$

14) $7 + 6x - 11 - 5x =$

15) $27x + 8 - 13 - 5x =$

16) $(-11)(-5x + 2) - 41x =$

17) $19x - 4(4 - 2x) =$

18) $16x + 3(3x + 6) + 10 =$

19) $5(-2x - 4) - 13x =$

20) $16x - 3x(x + 10) =$

21) $17x + 5x(2 - 4x) =$

22) $5x(-4x - 7) + 20x =$

23) $25x - 19 + 4x^2 =$

24) $6x(x - 11) + 25 =$

25) $4x - 5 + 15x + 3x^2 =$

26) $-7x^2 - 11x - 9x =$

27) $10x - 9x^2 - 3x^2 - 7 =$

28) $13 + 3x^2 - 9x^2 - 21x =$

29) $22x + 10x^2 - 15x + 17 =$

30) $4x^2 + 25x + 21x^2 =$

31) $29 - 12x^2 - 23x - 4x^2 =$

32) $22x - 19x - 9x^2 + 30 =$

WWW.MathNotion.Com

Common Core Subject Test Mathematics Grade 6

The Distributive Property

✏️ Use the distributive property to simply each expression.

1) $4(1 + 2x) =$

2) $2(4 + 7x) =$

3) $3(4x - 4) =$

4) $(2x - 5)(-6) =$

5) $(-3)(x + 6) =$

6) $(4 + 3x)2 =$

7) $(-5)(8 - 3x) =$

8) $-(-5 - 7x) =$

9) $(-6x + 3)(-3) =$

10) $(-4)(x - 7) =$

11) $-(5 - 3x) =$

12) $3(9 + 4x) =$

13) $6(4 + 3x) =$

14) $(-5x + 3)2 =$

15) $(5 - 8x)(-3) =$

16) $(-12)(3x + 3) =$

17) $(5 - 3x)6 =$

18) $4(2 + 6x) =$

19) $8(7x - 3) =$

20) $(-2x + 3)4 =$

21) $(7 - 5x)(-9) =$

22) $(-10)(x - 8) =$

23) $(11 - 4x)3 =$

24) $(-6)(10x - 4) =$

25) $(3 - 9x)(-7) =$

26) $(-9)(x + 9) =$

27) $(-3 + 5x)(-7) =$

28) $(-5)(8 - 10x) =$

29) $12(4x - 8) =$

30) $(-10x + 13)(-3) =$

31) $(-8)(3x - 2) + 4(x + 5) =$

32) $(-8)(x + 4) - (6 + 5x) =$

WWW.MathNotion.Com

Common Core Subject Test Mathematics Grade 6

Evaluating One Variable Expressions

✎ **Evaluate each expression using the value given.**

1) $8 - x, x = 5$

2) $x - 9, x = 5$

3) $5x + 4, x = 3$

4) $x - 13, x = -4$

5) $12 - x, x = 4$

6) $x + 2, x = 6$

7) $4x + 8, x = 3$

8) $x + (-7), x = -8$

9) $4x + 5, x = 2$

10) $3x + 9, x = -2$

11) $15 + 3x - 7, x = 2$

12) $17 - 3x, x = 3$

13) $8x - 9, x = 4$

14) $5x + 4, x = -3$

15) $10x + 5, x = 3$

16) $14 - 4x, x = -6$

17) $3(5x + 3), x = 9$

18) $4(-3x - 6), x = 3$

19) $7x - 2x + 12, x = 4$

20) $(5x + 6) \div 2, x = 8$

21) $(x + 18) \div 10, x = 12$

22) $5x - 12 + 3x, x = -3$

23) $(6 - 4x)(-3), x = -4$

24) $9x^2 + 3x - 6, x = 2$

25) $x^2 - 10x, x = -5$

26) $3x(7 - 2x), x = 2$

27) $12x + 6 - 2x^2, x = -4$

28) $(-3)(4x - 8 + 3x), x = 3$

29) $(-6) + \frac{x}{4} + 3x, x = 16$

30) $(-6) + \frac{x}{5}, x = 35$

31) $\left(-\frac{45}{x}\right) - 7 + 2x, x = 9$

32) $\left(-\frac{21}{x}\right) - 12 + 4x, x = 7$

Common Core Subject Test Mathematics Grade 6

Combining like Terms

✎ Simplify each expression.

1) $11x + 3x + 6 =$

2) $8(2x - 6) =$

3) $18x - 7x + 11 =$

4) $(-4)(6x - 7) =$

5) $22x - 10x - 5 =$

6) $32x - 13 + 8x =$

7) $15 - (8x - 11) =$

8) $-24x + 17 - 11x =$

9) $12x - 8 - 6x + 9 =$

10) $21x + 5 - 36 + 12x =$

11) $28x + 3x - 11 =$

12) $(-3x + 4)5 =$

13) $2 + 4x + 9x - 8 =$

14) $6(2x - 5x) - 4 =$

15) $4(5x + 11) + 3x =$

16) $x - 14 - 11x =$

17) $5(10 + 9x) - 8x =$

18) $42x + 17 - 23x =$

19) $(-7x) + 19 + 20x =$

20) $(-7x) - 33 + 29x =$

21) $4(5x + 3) - 19x =$

22) $5(6 - 2x) - 15x =$

23) $-24x + (11 - 18x) =$

24) $(-9) - (6)(7x + 3) =$

25) $(-1)(8x - 10) - 21x =$

26) $-36x + 14 + 27x - 5x =$

27) $3(-13x + 6) - 17x =$

28) $-5x - 42 + 32x =$

29) $37x - 19x + 15 - 9x =$

30) $3(5x + 7x) - 31 =$

31) $14 - 6x - 15 - 9x =$

32) $-2(-5x - 7x) + 27x =$

WWW.MathNotion.Com

Common Core Subject Test Mathematics Grade 6

Answers of Worksheets

Find a rule.

1)

Input	x	15	18	27	32.25	48.5
Output	y	4.5	7.5	16.5	21.75	38

2)

Input	x	3	9	15	21	33
Output	y	16	48	80	112	176

3)

Input	x	513	387	342	198	126
Output	y	57	43	38	22	14

4) $y = 2.5x$ 5) $y = x + 9.4$ 6) $y = x \div 8$

Translate Phrases into an Algebraic Statement

1) $9x$
2) $y - 11$
3) $\frac{19}{x}$
4) $38 - y$
5) $y + 40$
6) 6^2
7) x^5
8) $6 + x$
9) $57 - y$
10) $\frac{9}{x}$
11) $\frac{x^2}{25}$
12) $x - 6 = 19$
13) $10a - b^2$
14) $41 - ab$

Simplifying Variable Expressions

1) $3x + 15$
2) $-28x + 20$
3) $5x + 5$
4) $-8x^2 - 4$
5) $13x^2 + 10$
6) $18x^2 + 7x$
7) $-9x^2 + 4x$
8) $4x^2 - 10x$
9) $-22x + 21$
10) $68x - 12$
11) $-18x - 71$
12) $-11x^2 + 5x$
13) $-6x + 12$
14) $x - 4$
15) $22x - 5$
16) $14x - 22$
17) $27x - 16$
18) $25x + 28$
19) $-23x - 20$
20) $-3x^2 - 14x$
21) $-20x^2 + 27x$
22) $-20x^2 - 15x$
23) $4x^2 + 25x - 19$
24) $6x^2 - 66x + 25$
25) $3x^2 + 19x - 5$
26) $-7x^2 - 20x$
27) $-12x^2 + 10x - 7$
28) $-6x^2 - 21x + 13$
29) $10x^2 + 7x + 17$
30) $25x^2 + 25x$
31) $-16x^2 - 23x + 29$
32) $-9x^2 + 3x + 30$

The Distributive Property

1) $8x + 4$
2) $14x + 8$
3) $12x - 12$
4) $-12x + 30$
5) $-3x - 18$
6) $6x + 8$
7) $15x - 40$
8) $7x + 5$

WWW.MathNotion.Com

Common Core Subject Test Mathematics Grade 6

9) $18x - 9$
10) $-4x + 28$
11) $3x - 5$
12) $12x + 27$
13) $18x + 24$
14) $-10x + 6$
15) $24x - 15$
16) $-36x - 36$
17) $-18x + 30$
18) $24x + 8$
19) $56x - 24$
20) $-8x + 12$
21) $45x - 63$
22) $-10x + 80$
23) $-12x + 33$
24) $-60x + 24$
25) $63x - 21$
26) $-9x - 81$
27) $-35x + 21$
28) $50x - 40$
29) $48x - 96$
30) $30x - 39$
31) $-20x + 36$
32) $-13x - 38$

Evaluating One Variables

1) 3
2) -4
3) 19
4) -17
5) 8
6) 8
7) 20
8) -15
9) 13
10) 3
11) 14
12) 8
13) 23
14) -11
15) 35
16) 38
17) 144
18) -60
19) 32
20) 23
21) 3
22) -36
23) -66
24) 36
25) 75
26) 18
27) -74
28) -39
29) 46
30) 1
31) 6
32) 13

Combining like Terms

1) $14x + 6$
2) $16x - 48$
3) $11x + 11$
4) $-24x + 28$
5) $12x - 5$
6) $40x - 13$
7) $-8x + 26$
8) $-35x + 17$
9) $6x + 1$
10) $33x - 31$
11) $31x - 11$
12) $-15x + 20$
13) $13x - 6$
14) $-18x - 4$
15) $23x + 44$
16) $-10x - 14$
17) $37x + 50$
18) $19x + 17$
19) $13x + 19$
20) $22x - 33$
21) $x + 12$
22) $-25x + 30$
23) $-42x + 11$
24) $-42x - 27$
25) $-29x + 10$
26) $-14x + 14$
27) $-56x + 18$
28) $27x - 42$
29) $9x + 15$
30) $36x - 31$
31) $-15x - 1$
32) $51x$

WWW.MathNotion.Com

Common Core Subject Test Mathematics Grade 6

Chapter 9:
Equations and Inequalities

Topics that you will practice in this chapter:

- ✓ One-Step Equations
- ✓ Two-Step Equations
- ✓ Multi-Step Equations
- ✓ Graphing Single-Variable Inequalities
- ✓ One-Step Inequalities
- ✓ Two-Step Inequalities
- ✓ Multi-Step Inequalities

"Life is a math equation. In order to gain the most, you have to know how to convert negatives into positives." – Anonymous

Common Core Subject Test Mathematics Grade 6

One–Step Equations

✎ Find the answer for each equation.

1) $3x = 90, x =$ ____

2) $5x = 35, x =$ ____

3) $6x = 24, x =$ ____

4) $24x = 144, x =$ ____

5) $x + 15 = 20, x =$ ____

6) $x - 7 = 4, x =$ ____

7) $x - 9 = 2, x =$ ____

8) $x + 15 = 23, x =$ ____

9) $x - 4 = 13, x =$ ____

10) $12 = 16 + x, x =$ ____

11) $x - 10 = 2, x =$ ____

12) $5 - x = -11, x =$ ____

13) $28 = -6 + x, x =$ ____

14) $x - 20 = -35, x =$ ____

15) $x + 14 = -4, x =$ ____

16) $14 = 28 - x, x =$ ____

17) $7 + x = -7, x =$ ____

18) $x - 16 = 4, x =$ ____

19) $30 = x - 15, x =$ ____

20) $x - 5 = -18, x =$ ____

21) $x - 10 = 24, x =$ ____

22) $x - 20 = -25, x =$ ____

23) $x - 17 = 30, x =$ ____

24) $-70 = x - 28, x =$ ____

25) $x - 9 = 13, x =$ ____

26) $36 = 4x, x =$ ____

27) $x - 35 = 25, x =$ ____

28) $x - 25 = 10, x =$ ____

29) $70 - x = 16, x =$ ____

30) $x - 10 = 14, x =$ ____

31) $17 - x = -13, x =$ ____

32) $x - 9 = -30, x =$ ____

Common Core Subject Test Mathematics Grade 6

One–Step Equation Word Problems

✎ Solve.

1) How many boxes of envelopes can you buy with $40 if one box costs $5?

2) After paying $8.15 for a salad, Riya has $61.53. How much money did she have before buying the salad?

3) How many packages of Tissues can you buy with $81 if one package costs $4.5?

4) Last week Joe ran 40 miles more than Harrison. Joe ran 78 miles. How many miles did Harrison run?

5) Last Friday Liam had $65.46. Over the weekend he received some money for cleaning the attic. He now has $60. How much money did he receive?

6) After paying $17.36 for a sandwich, Elise has $31.23. How much money did she have before buying the sandwich?

Common Core Subject Test Mathematics Grade 6

Two-Steps Equations

✎ **Solve each equation.**

1) $6(3 + x) = 42$

2) $(-7)(x - 2) = 56$

3) $(-8)(3x - 4) = (-16)$

4) $5(2 + x) = -15$

5) $19(3x + 11) = 38$

6) $4(2x + 2) = 24$

7) $5(8 + 3x) = (-20)$

8) $(-5)(5x - 3) = 40$

9) $2x + 12 = 16$

10) $\frac{4x - 5}{5} = 3$

11) $(-3) = \frac{x + 4}{7}$

12) $80 = (-8)(x - 3)$

13) $\frac{x}{3} + 7 = 19$

14) $\frac{1}{4} = \frac{1}{2} + \frac{x}{4}$

15) $\frac{11 + x}{5} = (-6)$

16) $(-3)(10 + 5x) = (-15)$

17) $(-3x) + 12 = 24$

18) $\frac{x + 5}{5} = -5$

19) $\frac{x + 23}{8} = 3$

20) $(-4) + \frac{x}{2} = (-14)$

21) $-5 = \frac{x + 7}{8}$

22) $\frac{9x - 3}{6} = 4$

23) $\frac{2x - 12}{8} = 6$

24) $40 = (-5)(x - 8)$

Common Core Subject Test Mathematics Grade 6

Multi–Step Equations

✎ **Find the answer for each equation.**

1) $3x + 3 = 9$

2) $-x + 5 = 12$

3) $4x - 8 = 8$

4) $-(3 - x) = 5$

5) $4x - 8 = 16$

6) $12x - 15 = 9$

7) $2x - 18 = 2$

8) $4x + 8 = 16$

9) $24x + 27 = 75$

10) $-14(3 + x) = 14$

11) $-3(2 + x) = 6$

12) $12 = -(x - 7)$

13) $3(3 - x) = 30$

14) $-15 = -(3x + 6)$

15) $40(3 + x) = 40$

16) $5(x - 10) = 25$

17) $-18 = x + 8x$

18) $3x + 25 = -2x - 10$

19) $7(6 + 3x) = -63$

20) $18 - 3x = -4 - 5x$

21) $4 - 6x = 36 + 2x$

22) $15 + 15x = -5 + 5x$

23) $42 = (-6x) - 7 + 7$

24) $21 = 3x - 21 + 4x$

25) $-18 = -6x - 9 + 3x$

26) $5x - 15 = -29 + 6x$

27) $7x - 18 = 4x + 3$

28) $-7 - 4x = 5(4 - x)$

29) $x - 5 = -5(-3 - x)$

30) $13x - 68 = 15x - 102$

31) $-5x - 3 = -3(9 + 3x)$

32) $-2x - 15 = 6x + 17$

Common Core Subject Test Mathematics Grade 6

One-Step Inequalities

✏ Solve each inequality.

1) $7x < 14$

2) $x + 7 \geq -8$

3) $x - 1 \leq 9$

4) $-2x + 4 > -10$

5) $x + 18 \geq -6$

6) $x + 9 \geq 5$

7) $x - \frac{1}{3} \leq 5$

8) $-7x < 42$

9) $-x + 8 > -3$

10) $\frac{x}{3} + 3 > -9$

11) $-x + 8 > -4$

12) $x - 14 \leq 18$

13) $-x - 5 \leq -7$

14) $x + 26 \geq -13$

15) $x + \frac{1}{3} \geq -\frac{2}{3}$

16) $x + 6 \geq -14$

17) $x - 42 \leq -48$

18) $x - 5 \leq 4$

19) $-x + 5 > -6$

20) $x + 6 \geq -12$

21) $8x + 6 \leq 22$

22) $4x - 3 \geq 9$

23) $3x - 5 < 22$

24) $6x - 8 \leq 40$

WWW.MathNotion.Com

Graphing Inequalities

✎ **Draw a graph for each inequality.**

1) $x > -1$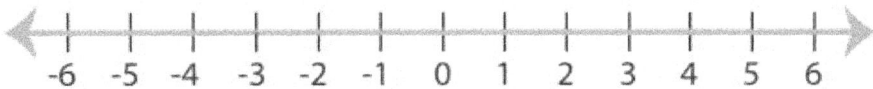

2) $x \leq 2$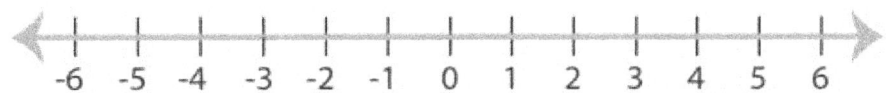

3) $x \geq 0$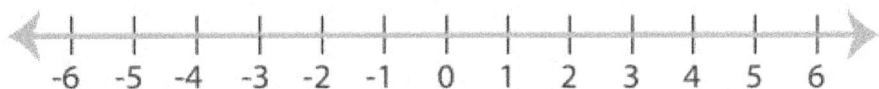

4) $x < -3$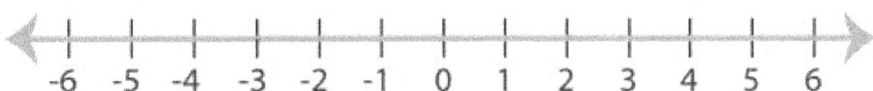

5) $x < \frac{1}{2}$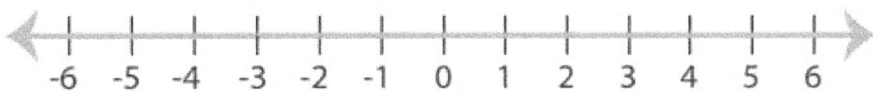

6) $x \leq -2$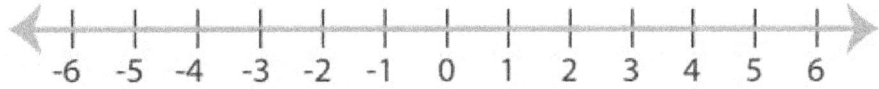

7) $x \leq 3$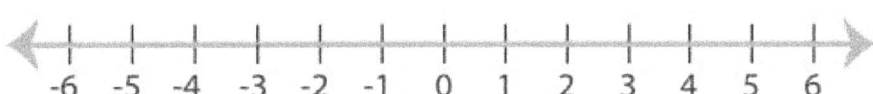

8) $x \geq -\frac{7}{2}$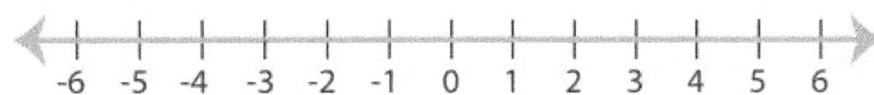

Common Core Subject Test Mathematics Grade 6

Two-Steps Inequality

👆 Solve each inequality

1) $2x - 3 \leq 7$

2) $3x - 4 \leq 8$

3) $\frac{-1}{4}x + \frac{x}{2} \leq \frac{1}{8}$

4) $5x + 10 \geq 30$

5) $4x - 7 \geq 9$

6) $3x - 5 \leq 16$

7) $8x - 2 \leq 14$

8) $9x + 5 \leq 23$

9) $2x + 10 > 32$

10) $\frac{x}{8} + 2 \leq 4$

11) $3x + 4 \geq 37$

12) $3x - 8 < 10$

13) $6 \geq \frac{x+7}{2}$

14) $3x + 9 < 48$

15) $\frac{4+x}{5} \geq 3$

16) $16 + 4x < 36$

17) $16 > 6x - 8$

18) $5 + \frac{x}{3} < 6$

19) $-4 + 4x > 24$

20) $5 + \frac{x}{7} < 3$

WWW.MathNotion.Com

Multi-Step Inequalities

✏️ **Calculate each inequality.**

21) $x - 3 \leq 7$

22) $8 - x \leq 8$

23) $3x - 9 \leq 9$

24) $4x - 4 \geq 8$

25) $x - 7 \geq 1$

26) $5x - 15 \leq 5$

27) $6x - 8 \leq 4$

28) $-11 + 6x \leq 12$

29) $4(x - 4) \leq 16$

30) $3x - 10 \leq 11$

31) $5x - 25 < 25$

32) $9x - 5 < 22$

33) $20 - 7x \geq -15$

34) $33 + 6x < 45$

35) $8 + 8x \geq 96$

36) $7 + 3x < 13$

37) $4x - 3 < 9$

38) $5(2 - 2x) \geq -30$

39) $-(7 + 6x) < 29$

40) $12 - 8x \geq -20$

41) $-4(x - 6) > 24$

42) $\dfrac{3x + 9}{6} \leq 10$

43) $\dfrac{4x - 10}{3} \leq 2$

44) $\dfrac{2x - 8}{3} > 2$

45) $8 + \dfrac{x}{6} < 9$

46) $\dfrac{9x}{7} - 4 < 5$

47) $\dfrac{15x + 45}{15} > 1$

48) $16 + \dfrac{x}{4} < 6$

Common Core Subject Test Mathematics Grade 6

Answers of Worksheets

One–Step Equations

1) 30
2) 7
3) 4
4) 6
5) 5
6) 11
7) 11
8) 8
9) 17
10) −4
11) 12
12) 16
13) 34
14) −15
15) −18
16) 14
17) −14
18) 20
19) 45
20) −13
21) 34
22) −5
23) 47
24) −42
25) 22
26) 9
27) 60
28) 35
29) 54
30) 24
31) 30
32) −21

One–Step Equation Word Problems

1) 8
2) $69.68
3) 18
4) 38
5) 5.46
6) 48.59

Two Steps Equations

1) $x = 4$
2) $x = -6$
3) $x = 2$
4) $x = -5$
5) $x = -3$
6) $x = 2$
7) $x = -4$
8) $x = -1$
9) $x = 2$
10) $x = 5$
11) $x = -25$
12) $x = -7$
13) $x = 36$
14) $x = -1$
15) $x = -41$
16) $x = -1$
17) $x = -4$
18) $x = -30$
19) $x = 1$
20) $x = -20$
21) $x = -47$
22) $x = 3$
23) $x = 30$
24) $x = 0$

Multi–Step Equations

1) 2
2) −7
3) 4
4) 8
5) 6
6) 2
7) 10
8) 2
9) 2
10) −4
11) −4
12) −5
13) −7
14) 3
15) −2
16) 15
17) −2
18) −7
19) −5
20) −11
21) −4
22) −2
23) −7
24) 6
25) 3
26) 14
27) 7
28) 27

WWW.MathNotion.Com

Common Core Subject Test Mathematics Grade 6

29) -5 30) 17 31) -6 32) -4

One Step Inequality

1) $x < 2$
2) $x \geq -15$
3) $x \leq 10$
4) $x < 7$
5) $x \geq -24$
6) $x \geq -4$
7) $x \leq \frac{16}{3}$
8) $x > -6$
9) $x < 11$
10) $x > -36$
11) $x < 12$
12) $x \leq 32$
13) $x \geq 2$
14) $x \geq -39$
15) $x \geq -1$
16) $x \geq -20$
17) $x \leq -6$
18) $x \leq 9$
19) $x < 11$
20) $x \geq -18$
21) $x \leq 2$
22) $x \geq 3$
23) $x < 9$
24) $x \leq 8$

Graphing Single–Variable Inequalities

1)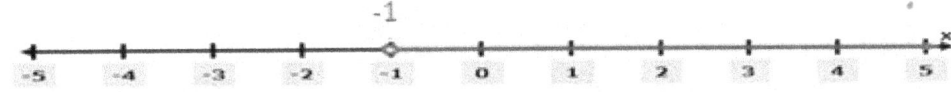

2)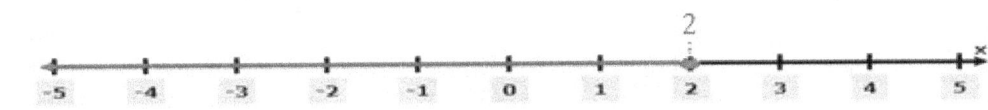

3)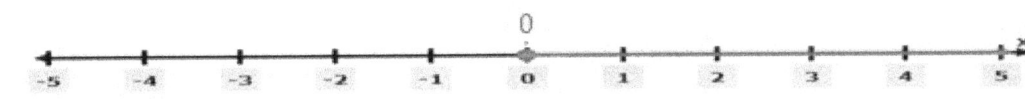

4)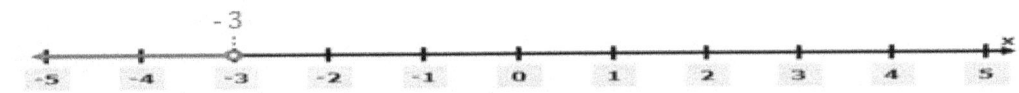

5)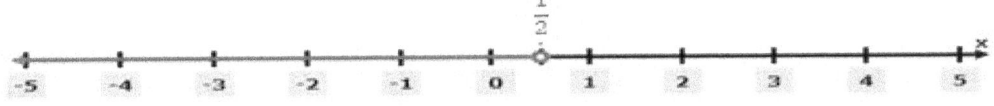

6)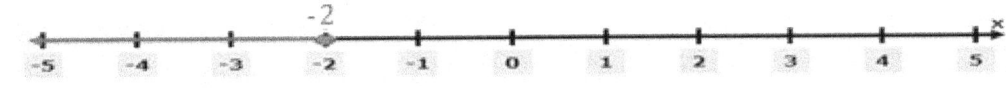

7)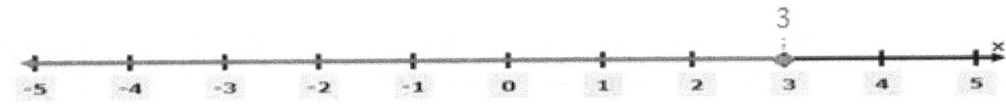

8)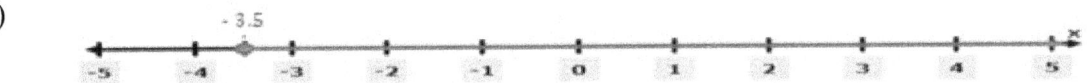

Two Steps Inequality

1) $x \leq 5$
2) $x \leq 4$
3) $x \leq 0.5$
4) $x \geq 4$
5) $x \geq 4$
6) $x \leq 7$

WWW.MathNotion.Com

Common Core Subject Test Mathematics Grade 6

7) $x \leq 2$
8) $x \leq 2$
9) $x > 11$
10) $x \leq 16$
11) $x \geq 11$

12) $x < 6$
13) $x \leq 5$
14) $x < 13$
15) $x \geq 11$
16) $x < 5$

17) $x < 4$
18) $x < 3$
19) $x > 7$
20) $x < -14$

Multi-Step Inequalities

1) $x \leq 10$
2) $x \geq 0$
3) $x \leq 6$
4) $x \geq 3$
5) $x \geq 8$
6) $x \leq 4$
7) $x \leq 2$

8) $x \leq \frac{23}{6}$
9) $x \leq 8$
10) $x \leq 7$
11) $x < 10$
12) $x < 3$
13) $x \leq 5$
14) $x < 2$

15) $x \geq 11$
16) $x < 2$
17) $x < 3$
18) $x \leq 4$
19) $x > -6$
20) $x \leq 4$
21) $x < 0$

22) $x \leq 17$
23) $x \leq 4$
24) $x > 7$
25) $x < 6$
26) $x < 7$
27) $x > -2$
28) $x < -40$

Common Core Subject Test Mathematics Grade 6

Chapter 10 :
Geometry and Solid Figures

Topics that you will practice in this chapter:

- ✓ Angles
- ✓ Pythagorean Relationship
- ✓ Triangles
- ✓ Polygons
- ✓ Trapezoids
- ✓ Circles
- ✓ Cubes
- ✓ Rectangular Prism
- ✓ Cylinder

Mathematics is, as it were, a sensuous logic, and relates to philosophy as do the arts, music, and plastic art to poetry. — *K. Shegel*

Common Core Subject Test Mathematics Grade 6

Angles

✎ **What is the value of *x* in the following figures?**

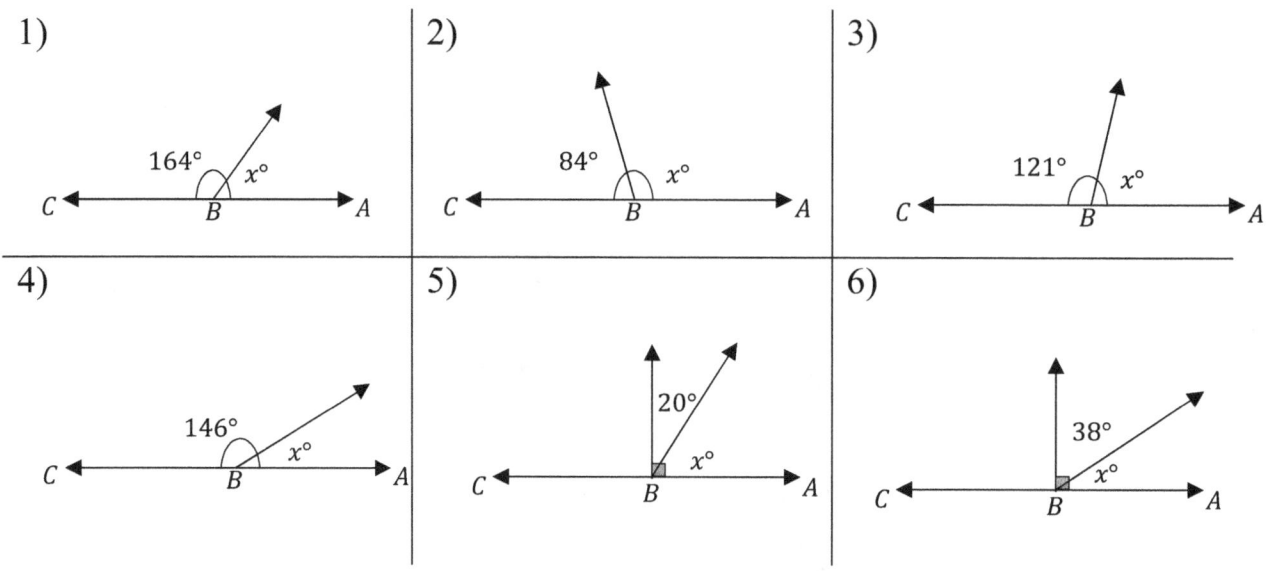

✎ **Calculate.**

7) Two supplement angles have equal measures. What is the measure of each angle? _____

8) The measure of an angle is seven fifth the measure of its supplement. What is the measure of the angle? _____

9) Two angles are complementary and the measure of one angle is 24 less than the other. What is the measure of the smaller angle? _____

10) Two angles are complementary. The measure of one angle is one fifth the measure of the other. What is the measure of the bigger angle? _____

11) Two supplementary angles are given. The measure of one angle is 40° less than the measure of the other. What does the smaller angle measure? _____

Common Core Subject Test Mathematics Grade 6

Pythagorean Relationship

✎ **Do the following lengths form a right triangle?**

1)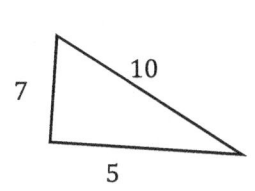

2)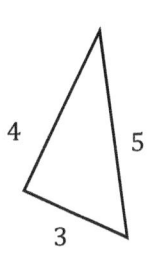

3)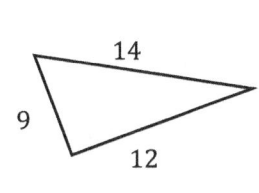

4)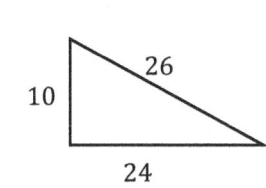

5)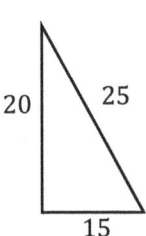

6)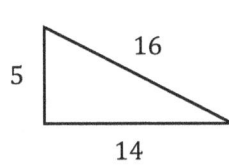

7)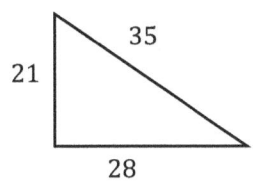

8)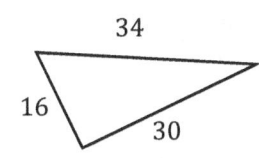

✎ **Find the missing side?**

9)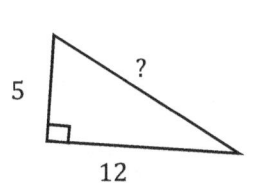

10)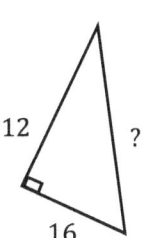

11)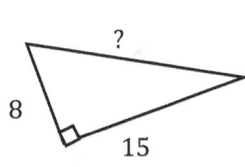

12)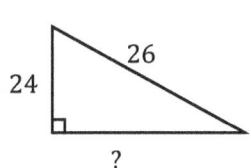

13)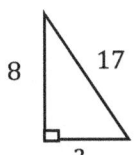

14)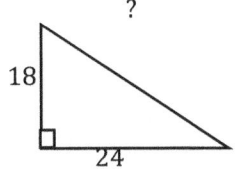

15)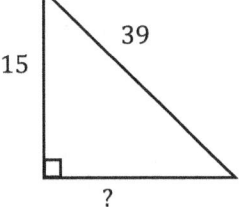

16)

WWW.MathNotion.Com

Common Core Subject Test Mathematics Grade 6

Triangles

✎ **Find the measure of the unknown angle in each triangle.**

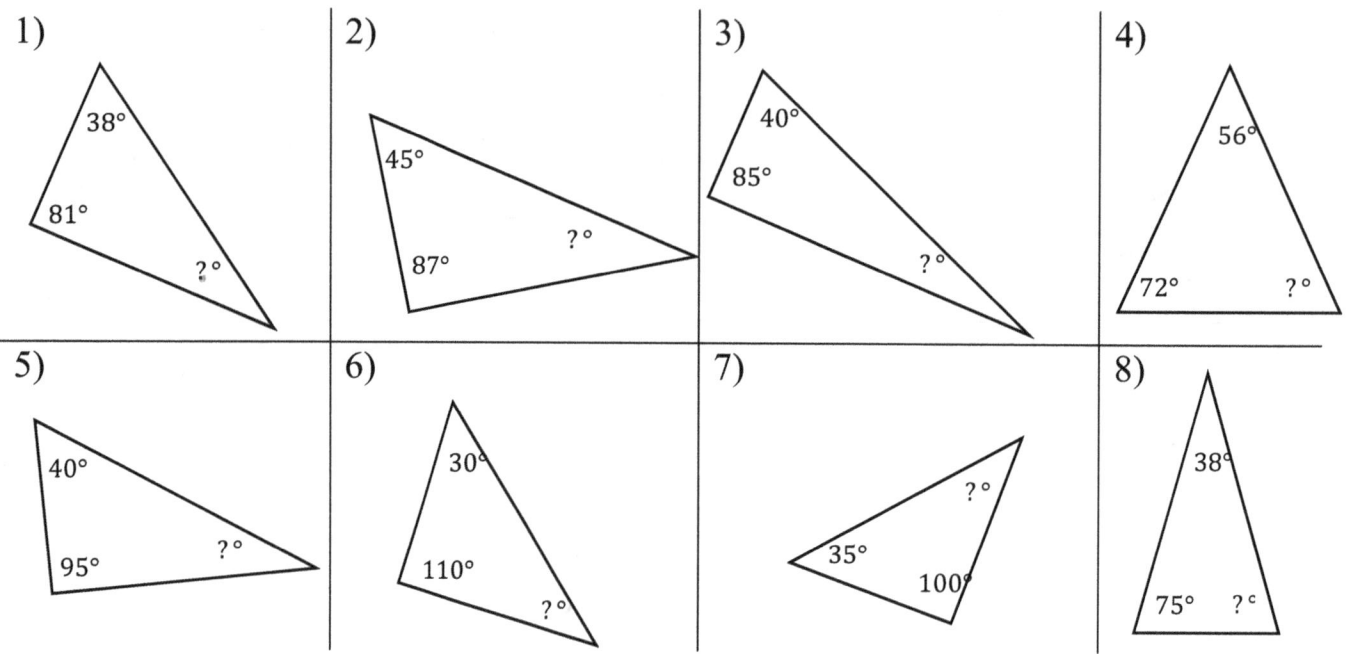

✎ **Find area of each triangle.**

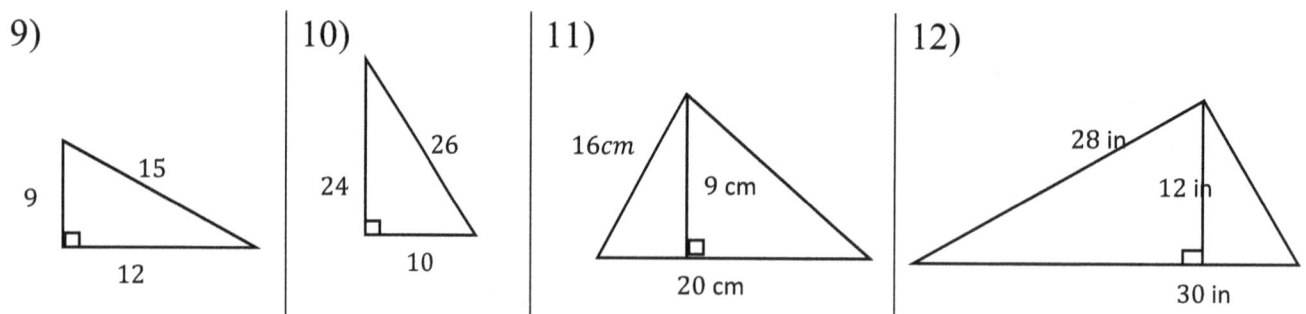

WWW.MathNotion.Com

Polygons

✎ **Find the perimeter of each shape.**

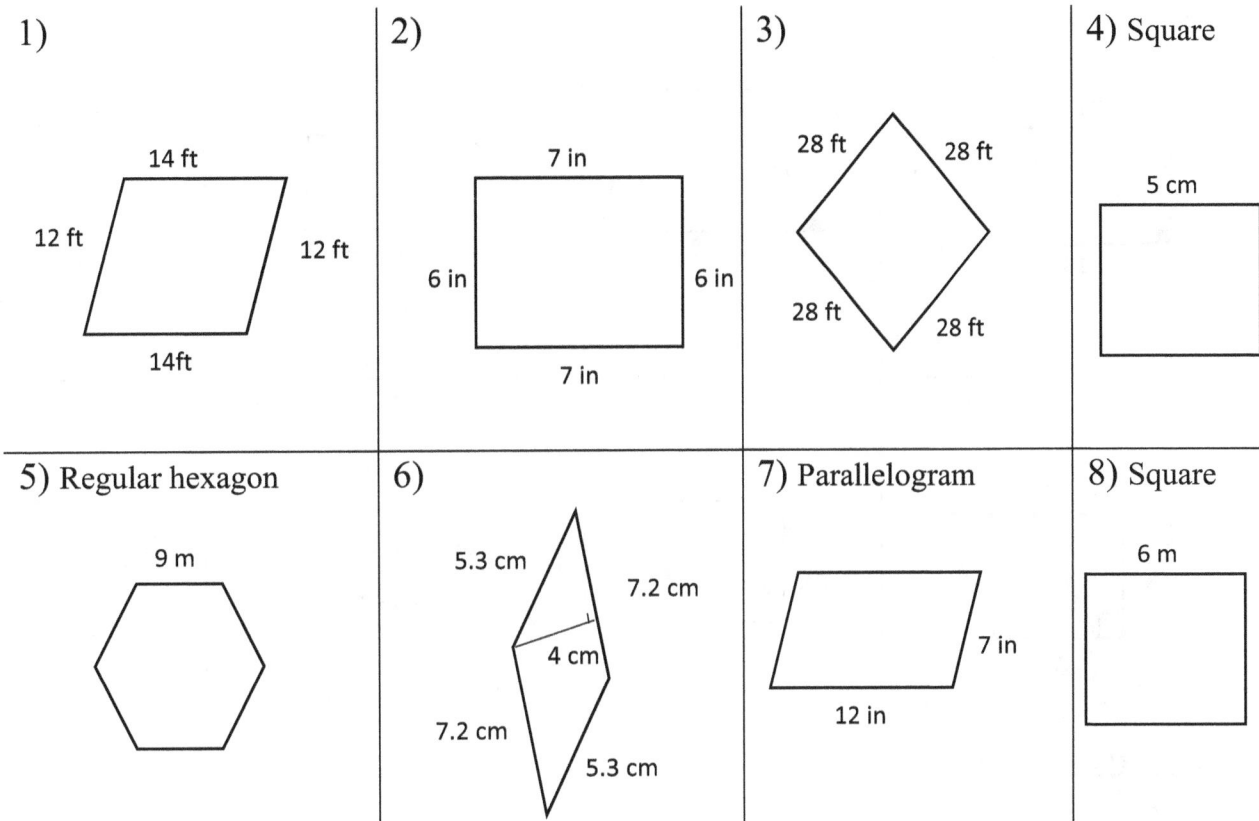

✎ **Find the area of each shape.**

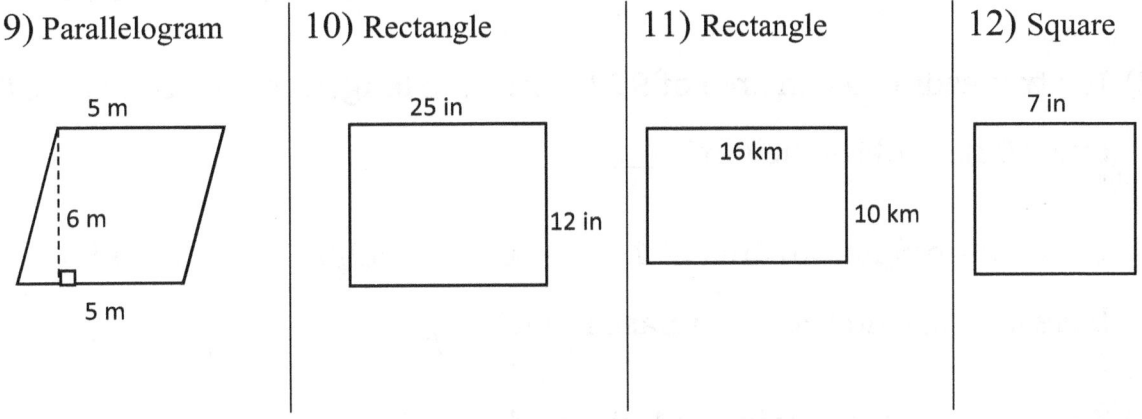

Common Core Subject Test Mathematics Grade 6

Trapezoids

✎ Find the area of each trapezoid.

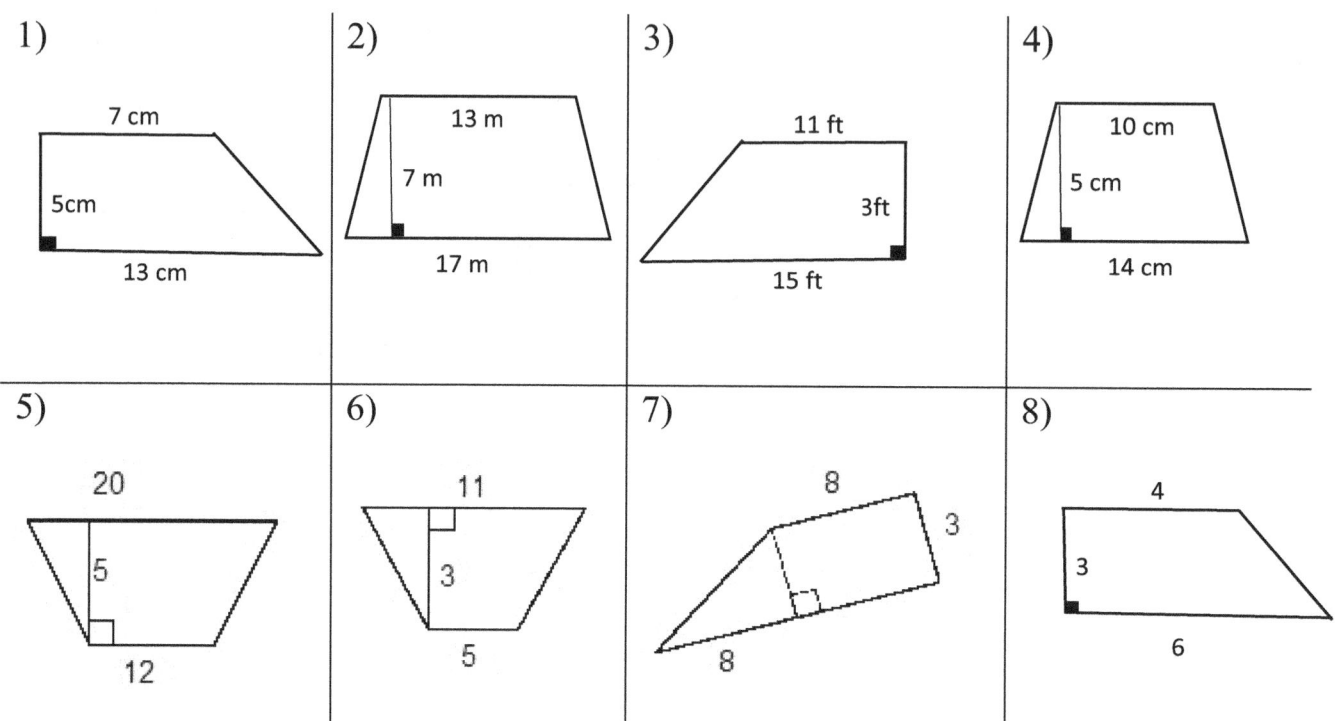

✎ Calculate.

1) A trapezoid has an area of 45 cm² and its height is 5 cm and one base is 5 cm. What is the other base length? _____

2) If a trapezoid has an area of 99 ft² and the lengths of the bases are 8 ft and 10 ft, find the height? _____

3) If a trapezoid has an area of 126 m² and its height is 14 m and one base is 6 m, find the other base length? _____

4) The area of a trapezoid is 440 ft² and its height is 22 ft. If one base of the trapezoid is 15 ft, what is the other base length?
_____

WWW.MathNotion.Com

Common Core Subject Test Mathematics Grade 6

Circles

✎ **Find the area of each circle.** ($\pi = 3.14$)

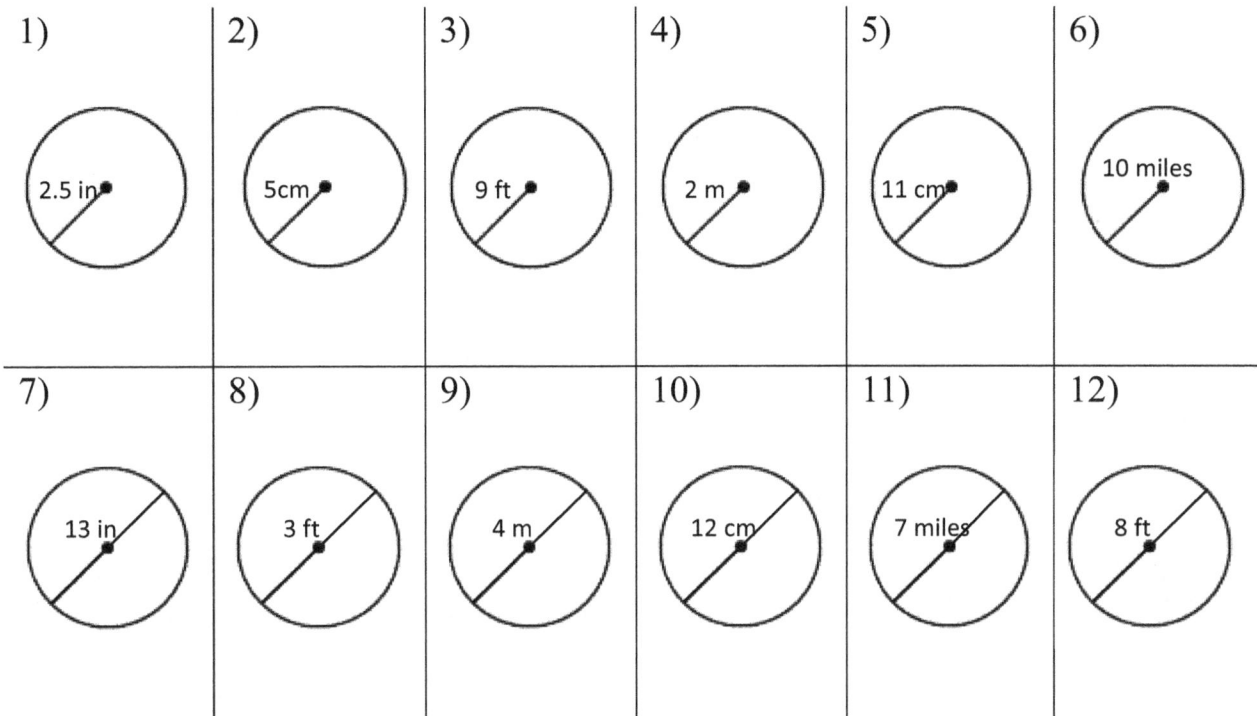

✎ **Complete the table below.** ($\pi = 3.14$)

Circle No.	Radius	Diameter	Circumference	Area
1	1 in	2 in	6.28 in	3.14 in^2
2		10 m		
3				28.26 ft^2
4			47.1 mi	
5		11 km		
6	7 cm			
7		12 ft		
8				314 m^2
9			56.52 in	
10	4.5 ft			

WWW.MathNotion.Com

Common Core Subject Test Mathematics Grade 6

Cubes

✎ Find the volume of each cube.

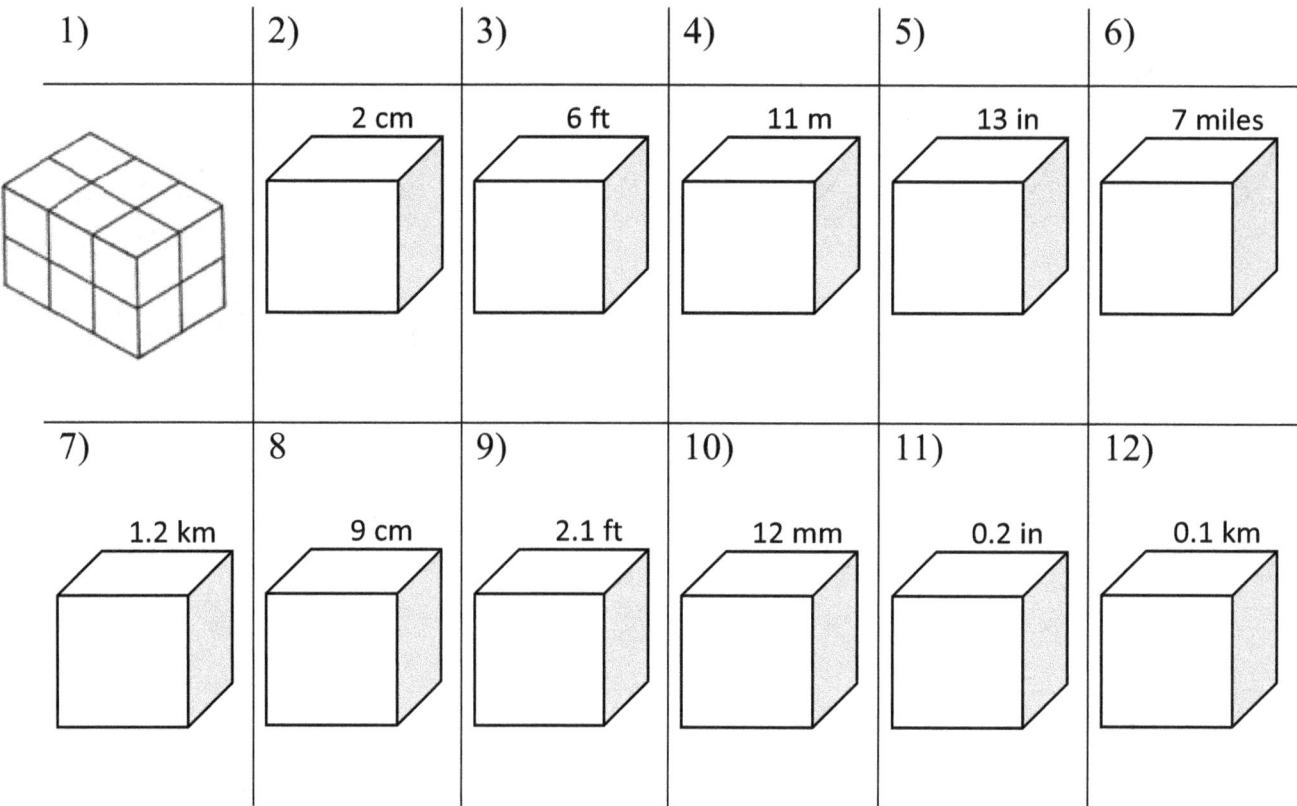

✎ Find the surface area of each cube.

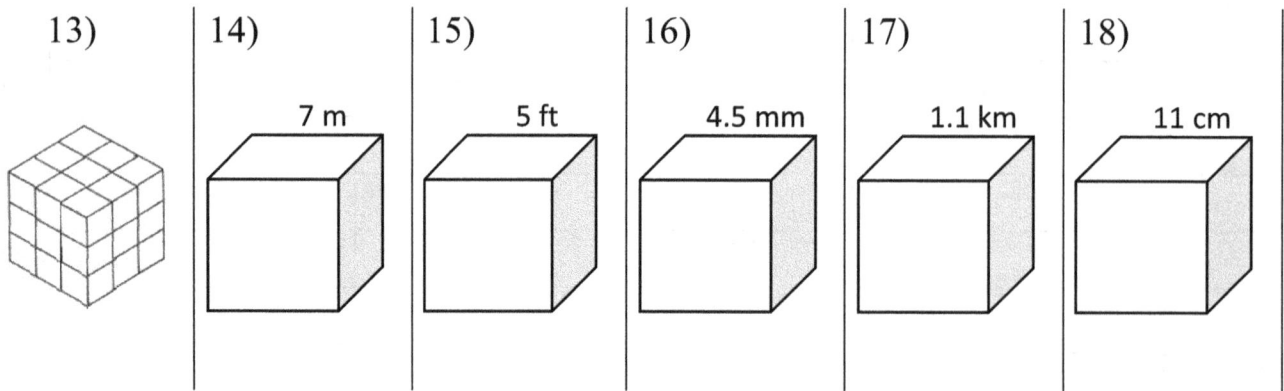

Rectangular Prism

✎ Find the volume of each Rectangular Prism.

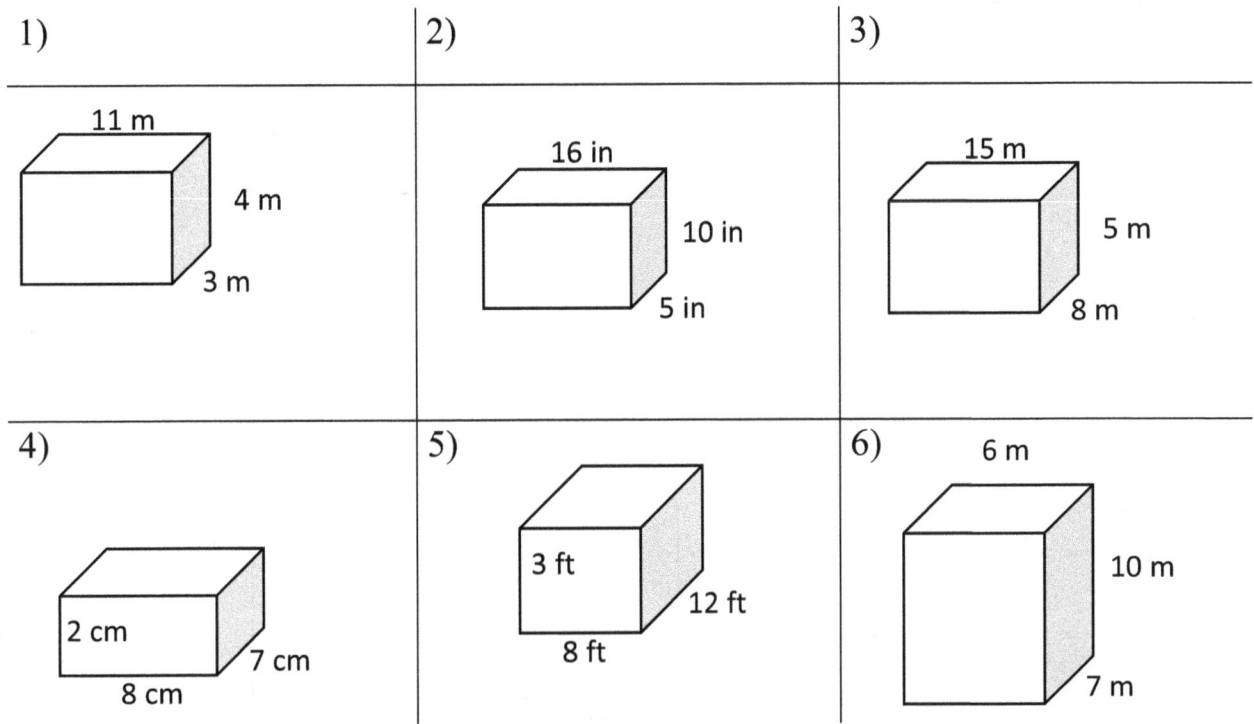

✎ Find the surface area of each Rectangular Prism.

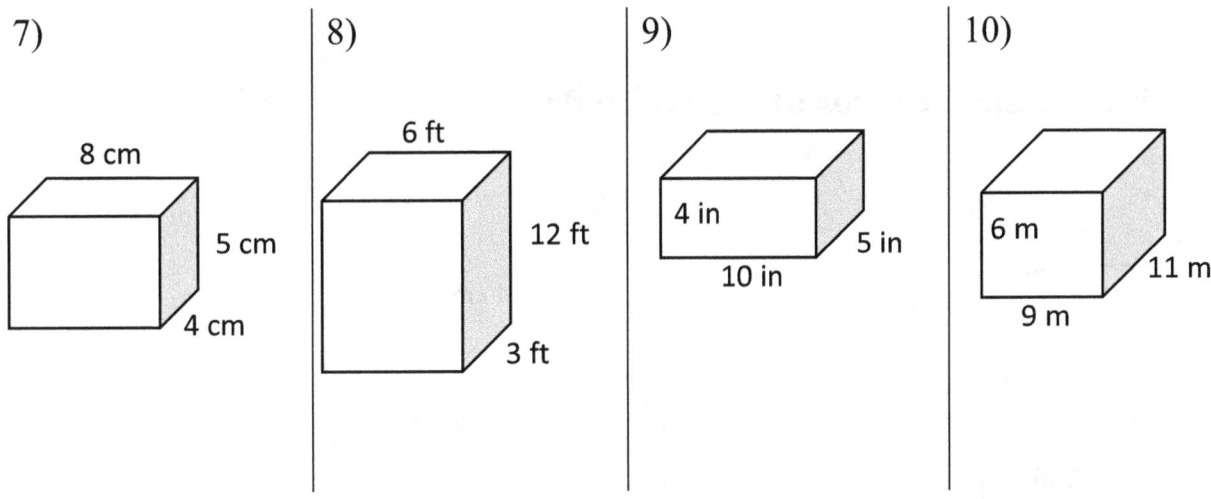

Common Core Subject Test Mathematics Grade 6

Cylinder

✏️ **Find the volume of each Cylinder. Round your answer to the nearest tenth.** ($\pi = 3.14$)

1)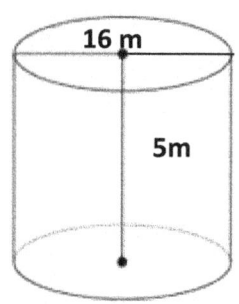

2)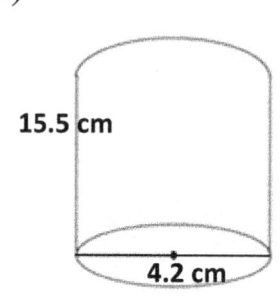

3)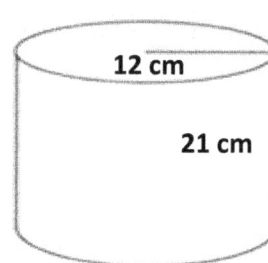

4)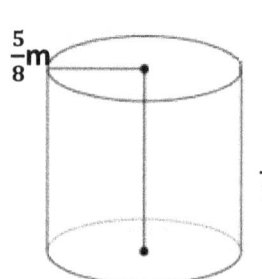

5)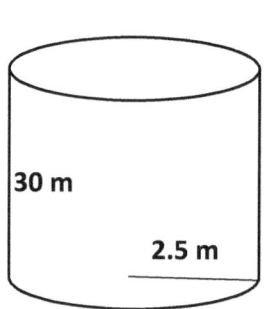

6)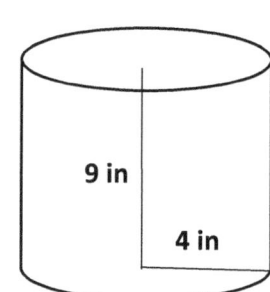

✏️ **Find the surface area of each Cylinder.** ($\pi = 3.14$)

7)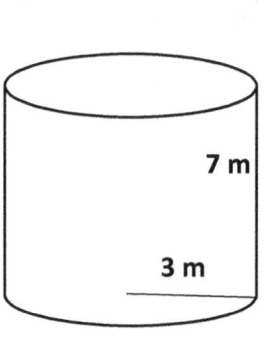

8)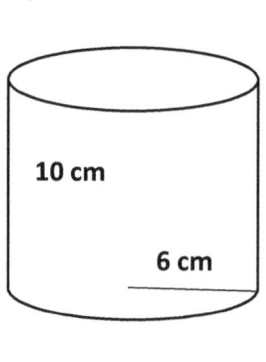

9)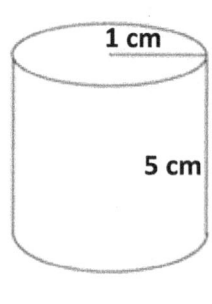

10)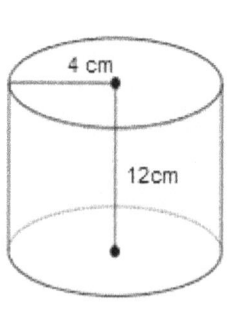

WWW.MathNotion.Com

Common Core Subject Test Mathematics Grade 6

Answers of Worksheets

Angles

1) 16°   4) 34°   7) 90°   10) 75°
2) 96°   5) 70°   8) 75°   11) 70°
3) 59°   6) 52°   9) 33°

Pythagorean Relationship

1) No   5) Yes   9) 13   13) 15
2) Yes   6) No   10) 20   14) 30
3) No   7) Yes   11) 17   15) 36
4) Yes   8) Yes   12) 10   16) 12

Triangles

1) 60°   5) 45°   9) 54 square unites
2) 48°   6) 40°   10) 120 square unites
3) 55°   7) 45°   11) 90 square unites
4) 52°   8) 67°   12) 180 square unites

Polygons

1) 52 ft   5) 54 m   9) 30 m^2
2) 26 in   6) 25 cm   10) 300 in^2
3) 112 ft   7) 38 in   11) 160 km^2
4) 20 cm   8) 24 m   12) 49 in^2

Trapezoids

1) 50 cm^2 4) 60 cm^2 7) 36
2) 105 m^2 5) 80   8) 15
3) 39 ft^2 6) 24

Calculate

1) 13 cm   2) 11 ft   3) 12 m   4) 25 ft

Circles

1) 19.63 in^2 5) 379.94 cm^2 9) 12.56 m^2
2) 78.5 cm^2 6) 314 $miles^2$ 10) 113.04 cm^2
3) 254.34 ft^2 7) 132.67 in^2 11) 38.47 $miles^2$
4) 12.56 m^2 8) 7.07 ft^2 12) 50.24 ft^2

WWW.MathNotion.Com

Common Core Subject Test Mathematics Grade 6

Circle No.	Radius	Diameter	Circumference	Area
1	1 in	2 in	6.28 in	3.14 in^2
2	5 m	10 m	31.4 m	78.5 m^2
3	3 ft	6 ft	18.84 ft	28.26 ft^2
4	7.5 miles	15 mi	47.1 mi	176.63 mi^2
5	5.5 km	11 km	34.54 km	94.99 km^2
6	7 cm	14 cm	43.96 cm	153.86 cm^2
7	6 ft	12 ft	37.68 feet	113.04 ft^2
8	10 m	20 m	62.8 m	314 m^2
9	9 in	18 in	56.52 in	254.34 in^2
10	4.5 ft	9 ft	28.26 ft	63.585 ft^2

Cubes

1) 12
2) 8 cm^3
3) 216 ft^3
4) 1,331 m^3
5) 2,197 in^3
6) 343 $miles^3$
7) 1.728 km^3
8) 729 cm^3
9) 9.261 ft^3
10) 1,728 mm^3
11) 0.008 in^3
12) 0.001 km^3
13) 27
14) 294 m^2
15) 150 ft^2
16) 121.5 mm^2
17) 7.26 km^2
18) 726 cm^2

Rectangular Prism

1) 132 m^3
2) 800 in^3
3) 600 m^3
4) 112 cm^3
5) 288 ft^3
6) 420 m^3
7) 184 cm^2
8) 252 ft^2
9) 220 in^2
10) 438 m^2

Cylinder

1) 1,004.8 m^3
2) 214.6 cm^3
3) 9,495.4 cm^3
4) 1.1 m^3
5) 588.8 m^3
6) 452.2 in^3
7) 188.4 m^2
8) 602.9 cm^2
9) 37.7 cm^2
10) 401.9 m^2

Common Core Subject Test Mathematics Grade 6

Chapter 11:
Statistics and Probability

Topics that you will practice in this chapter:

- ✓ Mean and Median
- ✓ Mode and Range
- ✓ Times Series
- ✓ Stem–and–Leaf Plot
- ✓ Pie Graph
- ✓ Probability Problems

Mathematics is no more computation than typing is literature.
– John Allen Paulos

Common Core Subject Test Mathematics Grade 6

Mean and Median

✎ **Find Mean and Median of the Given Data.**

1) 8, 7, 14, 4, 8

2) 14, 8, 25, 19, 16, 33, 11

3) 23, 18, 15, 12, 17

4) 34, 14, 10, 15, 6, 11

5) 10, 19, 6, 8, 32, 20, 17

6) 17, 26, 39, 69, 20, 6

7) 40, 38, 18, 11, 9, 2, 7, 32, 41

8) 24, 21, 31, 12, 33, 32, 22

9) 16, 14, 20, 41, 15, 20, 38, 4

10) 20, 20, 30, 18, 6, 28, 12, 46

11) 12, 7, 10, 11, 16, 22

12) 10, 29, 27, 12, 2, 15, 10, 3

✎ **Calculate.**

13) In a javelin throw competition, five athletics score 56, 34, 62, 23 and 19 meters. What are their Mean and Median? _____

14) Eva went to shop and bought 8 apples, 14 peaches, 6 bananas, 4 pineapples and 12 melons. What are the Mean and Median of her purchase? _____

15) Bob has 17 black pen, 19 red pen, 14 green pens, 20 blue pens and 5 boxes of yellow pens. If the Mean and Median are 19 respectively, what is the number of yellow pens in each box? _____

Common Core Subject Test Mathematics Grade 6

Mode and Range

✎ **Find Mode and Rage of the Given Data.**

1) 4, 3, 7, 3, 3, 4
   Mode: _____ Range: _____

2) 18, 18, 24, 26, 18, 8, 14, 22
   Mode: _____ Range: _____

3) 8, 8, 8, 16, 19, 22, 20, 9, 13
   Mode: _____ Range: _____

4) 24, 24, 14, 28, 20, 18, 20, 24
   Mode: _____ Range: _____

5) 6, 21, 27, 24, 27, 27
   Mode: _____ Range: _____

6) 21, 8, 8, 7, 8, 12, 10, 22, 18, 13
   Mode: _____ Range: _____

7) 7, 4, 4, 6, 13, 13, 13, 0, 2, 2
   Mode: _____ Range: _____

8) 5, 8, 5, 14, 12, 14, 3, 5, 18
   Mode: _____ Range: _____

9) 7, 7, 7, 12, 7, 3, 8, 16, 3, 17
   Mode: _____ Range: _____

10) 15, 15, 19, 16, 4, 16, 10, 15
    Mode: _____ Range: _____

11) 6, 6, 5, 6, 42, 13, 19, 2
    Mode: _____ Range: _____

12) 8, 8, 9, 8, 9, 4, 34, 22
    Mode: _____ Range: _____

✎ **Calculate.**

13) A stationery sold 12 pencils, 56 red pens, 24 blue pens, 20 notebooks, 12 erasers, 21 rulers and 11 color pencils. What are the Mode and Range for the stationery sells?

    Mode: _____ Range: _____

14) In an English test, eight students score 10, 15, 15, 18 18, 16, 15 and 15. What are their Mode and Range? _____

15) What is the range of the first 6 even numbers greater than 8?

    _____

Common Core Subject Test Mathematics Grade 6

Times Series

🖱 **Use the following Graph to complete the table.**

Day	Distance (km)
1	
2	

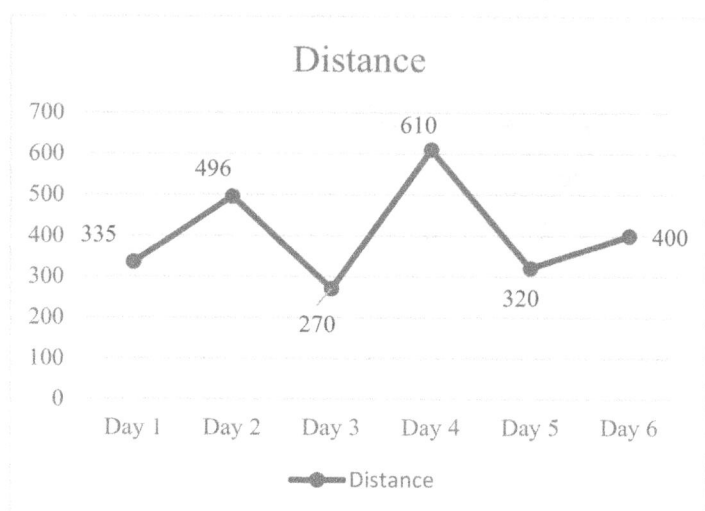

The following table shows the number of births in the US from 2007 to 2012 (in millions).

Year	Number of births (in millions)
2007	4.15
2008	3.70
2009	3.45
2010	3.20
2011	1.75
2012	2.98

Draw a Time Series for the table.

Common Core Subject Test Mathematics Grade 6

Stem–and–Leaf Plot

✎ **Make stem ad leaf plots for the given data.**

1) 24, 26, 29, 20, 53, 27, 51, 55, 36, 21, 37, 30

   Stem | Leaf plot

2) 11, 59, 66, 14, 18, 19, 59, 65, 69, 61, 68, 65

   Stem | Leaf plot

3) 121, 55, 66, 54, 112, 128, 63, 125, 59, 123, 68, 119

   Stem | Leaf plot

4) 51, 32, 100, 56, 84, 36, 107, 56, 85, 39, 56, 106, 89

   Stem | Leaf plot

5) 33, 89, 19, 87, 81, 16, 11, 30, 86, 35, 17, 35, 13

   Stem | Leaf plot

6) 60, 92, 22, 25, 67, 93, 95, 62, 21, 64, 98, 29

   Stem | Leaf plot

WWW.MathNotion.Com

Common Core Subject Test Mathematics Grade 6

Quartile of a Data Set

✍ **Find First, Second and Third Quartile of the Given Data.**

1) 45, 8, 25, 43, 24, 36, 35, 62, 19

2) 23, 63, 25, 19, 80, 32

3) 86, 33, 85, 60, 72, 42, 51, 46

4) 24, 44, 48, 25, 25, 36, 25, 36, 71, 49

5) 23, 15, 45, 9, 35, 8, 25, 15

6) 66, 86, 40, 32, 82, 25, 52, 44, 61

Box and Whisker Plots

✍ **Make box and whisker plots for the given data.**

1) 86, 65, 92, 67, 72, 87, 87, 83, 95, 66, 76, 82

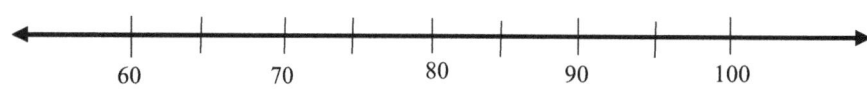

2) 8, 22, 17, 15, 13, 5, 8, 12, 6, 11, 6, 15, 4, 28

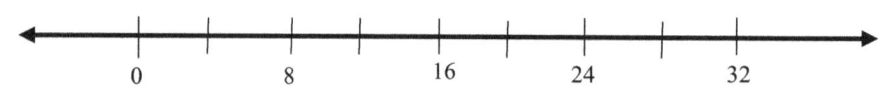

3) 25, 21, 34, 19, 23, 24, 13, 17, 15, 16, 22

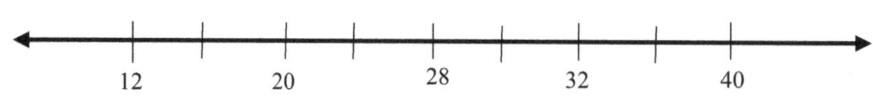

WWW.MathNotion.Com

Common Core Subject Test Mathematics Grade 6

Pie Graph

The circle graph below shows all Robert's expenses for last month. Robert spent $140 on his hobbies last month.

Answer following questions based on the Pie graph.

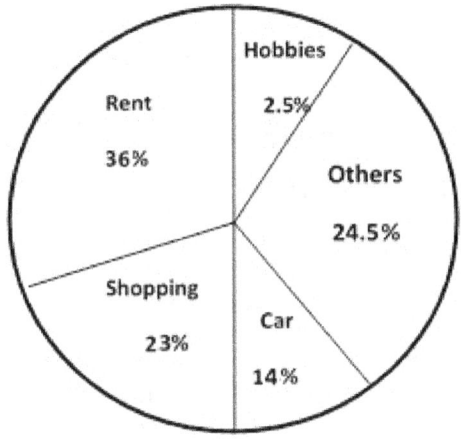

1) How much was Robert's total expenses last month? _____

2) How much did Robert spend on his car last month? _____

3) How much did Robert spend for shopping last month? _____

4) How much did Robert spend on his rent last month? _____

5) What fraction is Robert's expenses for his rent and car out of his total expenses last month? _____

Common Core Subject Test Mathematics Grade 6

Probability Problems

✎ **Calculate.**

1) A number is chosen at random from 1 to 10. Find the probability of selecting number 6 or smaller numbers. _____

2) Bag A contains 18 red marbles and 6 green marbles. Bag B contains 16 black marbles and 8 orange marbles. What is the probability of selecting a green marble at random from bag A? What is the probability of selecting a black marble at random from Bag B? _____

3) A number is chosen at random from 1 to 20. What is the probability of selecting multiples of 4? _____

4) A card is chosen from a well-shuffled deck of 52 cards. What is the probability that the card will be a queen? _____

5) A number is chosen at random from 1 to 15. What is the probability of selecting a multiple of 3 or 5? _____

A spinner numbered 1–8, is spun once. What is the probability of spinning …?

6) an Odd number? _____   7) a multiple of 2? _____

8) a multiple of 5? _____   9) number 10? _____

WWW.MathNotion.Com

Common Core Subject Test Mathematics Grade 6

Answers of Worksheets

Mean and Median

1) Mean: 8.2, Median: 8
2) Mean: 18, Median: 16
3) Mean: 17, Median: 17
4) Mean: 15, Median: 12.5
5) Mean: 16, Median: 17
6) Mean: 29.5, Median: 23
7) Mean: 22, Median: 18
8) Mean: 25, Median: 24
9) Mean: 21, Median: 18
10) Mean: 22.5, Median: 20
11) Mean: 13, Median: 11.5
12) Mean: 13.5, Median: 11
13) Mean: 38.8, Median: 34
14) Mean: 8.8, Median: 8
15) 5

Mode and Range

1) Mode: 3, Range: 4
2) Mode: 18, Range: 18
3) Mode: 8, Range: 14
4) Mode: 24, Range: 14
5) Mode: 27, Range: 21
6) Mode: 8, Range: 15
7) Mode: 13, Range: 13
8) Mode: 5, Range: 15
9) Mode: 7, Range: 14
10) Mode: 15, Range: 15
11) Mode: 6, Range: 40
12) Mode: 8, Range: 30
13) Mode: 12, Range: 45
14) Mode: 15, Range: 8
15) 10

Time series

Day	Distance (km)
1	335
2	496
3	270
4	610
5	320
6	400

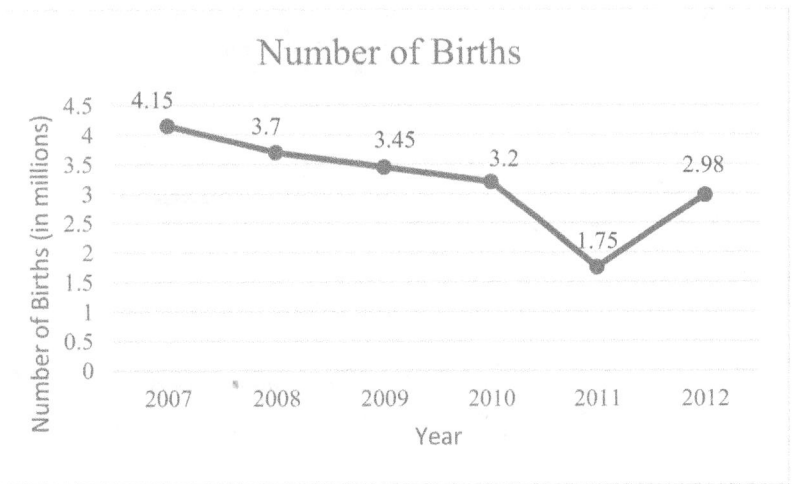

Stem–And–Leaf Plot

1)

Stem	leaf
2	0 1 4 6 7 9
3	0 6 7
5	1 3 5

2)

Stem	leaf
1	1 4 8 9
5	9 9
6	1 5 5 6 8 9

3)

Stem	leaf
5	4 5 9
6	3 6 8
11	2 9
12	1 3 5 8

Common Core Subject Test Mathematics Grade 6

4)

Stem	leaf
3	2 6 9
5	1 6 6 6
8	4 5 9
10	0 6 7

5)

Stem	leaf
1	1 3 6 7 9
3	0 3 5 5
8	1 6 7 9

6)

Stem	leaf
2	2 1 5 9
6	0 2 4 7
9	2 3 5 8

Quartile of the Given Data

1) First quartile: 21.5, second quartile: 35, third quartile: 44
2) First quartile: 22, second quartile: 28.5, third quartile: 67.25
3) First quartile: 43, second quartile: 55.5, third quartile: 81.75
4) First quartile: 25, second quartile: 36, third quartile: 48.25
5) First quartile: 10.5, second quartile: 19, third quartile: 32.5
6) First quartile: 36, second quartile: 52, third quartile: 74

Box and Whisker Plots

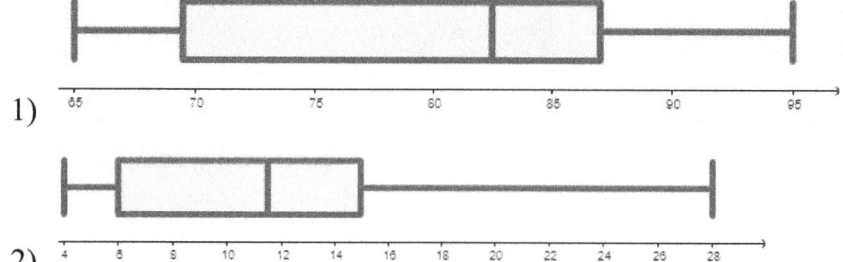

1)
2)

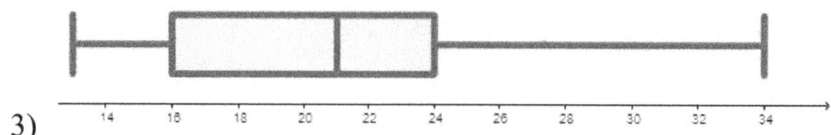

3)

Pie Graph

1) $5,600
2) $784
3) $1,288
4) $2,016
5) $\frac{1}{2}$

Probability Problems

1) $\frac{3}{5}$
2) $\frac{1}{4}, \frac{2}{3}$
3) $\frac{1}{4}$
4) $\frac{1}{13}$
5) $\frac{7}{15}$
6) $\frac{1}{2}$
7) $\frac{1}{2}$
8) $\frac{1}{8}$
9) 0

WWW.MathNotion.Com

Common Core Subject Test Mathematics Grade 6

Chapter 12 : Common Core Math Practice Tests

Time to Test

Time to refine your skill with a practice examination.

Take a REAL Common Core Mathematics test to simulate the test day experience. After you've finished, score your test using the answer key.

Before You Start

- You'll need a pencil and scratch papers to take the test.
- For this practice test, don't time yourself. Spend time as much as you need.
- It's okay to guess. You won't lose any points if you're wrong.
- After you've finished the test, review the answer key to see where you went wrong.

Calculators are not permitted for Common Core Tests.

Good Luck!

Common Core Subject Test Mathematics Grade 6

Common Core Subject Test Mathematics Grade 6

Common Core GRADE 6 MAHEMATICS REFRENCE

Conversions:

LENGTH

Customary	Metric
1 mile (mi) = 1,760 yards (yd)	1 kilometer (km) = 1,000 meters (m)
1 yard (yd) = 3 feet (ft)	1 meter (m) = 100 centimeters (cm)
1 foot (ft) = 12 inches (in.)	1 centimeter (cm) = 10 millimeters (mm)

VOLUME AND CAPACITY

Customary	Metric
1 gallon (gal) = 4 quarts (qt)	1 liter (L) = 1,000 milliliters (mL)
1 quart (qt) = 2 pints (pt.)	
1 pint (pt.) = 2 cups (c)	
1 cup (c) = 8 fluid ounces (Fl oz)	

WEIGHT AND MASS

Customary	Metric
1 ton (T) = 2,000 pounds (lb.)	1 kilogram (kg) = 1,000 grams (g)
1 pound (lb.) = 16 ounces (oz)	1 gram (g) = 1,000 milligrams (mg)

Formulas:

Area

Triangle	$A = \frac{1}{2}bh$
Rectangle or Parallelogram	$A = bh$
Trapezoid	$A = \frac{1}{2}h(b_1 + b_2)$

Volume

Rectangular Prism	$V = Bh$

MATERIALS

Common Core Subject Test Mathematics Grade 6

Common Core Subject Test Mathematics Grade 6

Common Core Practice Test 1

Mathematics

GRADE 6

Released *Month Year*

Common Core Subject Test Mathematics Grade 6

1) Martin earns $14 an hour. Which of the following inequalities represents the amount of time Martin needs to work per day to earn at least $165 per day?

   A. $14t \geq 165$

   B. $14t \leq 165$

   C. $14 + t \geq 165$

   D. $14 + t \leq 165$

2) What is the value of the expression $3(x - 3y) + (11 - 2x)^2$, when $x = 4$ and $y = -1$?

   A. -32

   B. 30

   C. 32

   D. -30

3) Round $\frac{178}{7}$ to the nearest tenth.

   A. 26.2

   B. 28.7

   C. 25.4

   D. 26.4

Common Core Subject Test Mathematics Grade 6

4) Which expression is equivalent to $8(7x - 11)$?

   A. -88

   B. $-56x$

   C. $88x - 56$

   D. $56x - 88$

5) A chemical solution contains 17% alcohol. If there is 85 ml of alcohol, what is the volume of the solution?

   A. 1,700 ml

   B. 500 ml

   C. 700 ml

   D. 1,300 ml

6) Which ordered pair describes point A that is shown below?

   A. $(-2, -2)$

   B. $(-2, 2)$

   C. $(2, -2)$

   D. $(2, 2)$

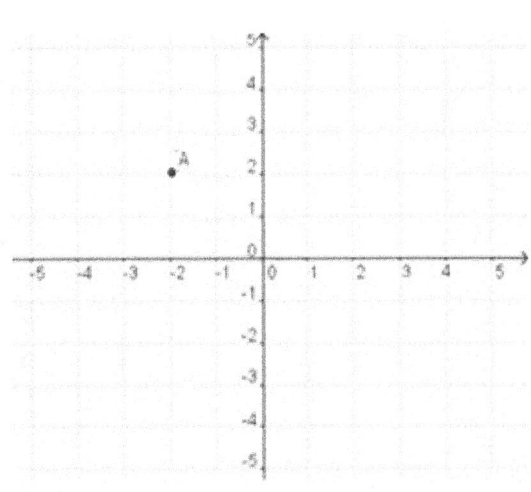

Common Core Subject Test Mathematics Grade 6

7) To produce a special concrete, for every 27 kg of cement, 9 liters of water is required. Which of the following ratios is the same as the ratio of cement to liters of water?

A. 108: 36

B. 94: 36

C. 108: 45

D. 51: 18

8) Find the opposite of the numbers 14, 6.

A. $\frac{1}{14}, 6$

B. $-14, 6$

C. $-14, -6$

D. $-\frac{1}{14}, -6$

9) Which of the following graphs represents the following inequality?

$$-3 \leq 7x - 3 < 11$$

A.

B.

C.

D.

Common Core Subject Test Mathematics Grade 6

10) What is the value of x in the following equation? $-18 = 41 - x$

   A. 55

   B. −55

   C. 59

   D. −59

11) The ratio of boys to girls in a school is 3:4. If there are 154 students in the school, how many boys are in the school?

   A. 22

   B. 66

   C. 88

   D. 120

12) $(64 + 8) \div 24$ is equivalent to …

   A. $64 \div 2.8$

   B. $\frac{64}{24} + 8$

   C. $(2 \times 2 \times 2 \times 3 \times 3) \div (2 \times 2 \times 2 \times 3)$

   D. $(2 \times 2 \times 2 \times 3 \times 3) \div 2 + 3$

13) What is the equation of a line that passes through points (4, 3) and (5, 8)?

   A. $y = -5x - 20$

   B. $y = 5x - 17$

   C. $y = 5x + 17$

   D. $y = -5x - 4$

Common Core Subject Test Mathematics Grade 6

14) What is the volume of a box with the following dimensions? Height = 3 cm

Width = 4 cm   Length = 11 cm

A. 132 cm^3

B. 144 cm^3

C. 236 cm^3

D. 240 cm^3

15) Anita's trick–or–treat bag contains 15 pieces of chocolate, 6 suckers, 12 pieces of gum, 17 pieces of licorice. If she randomly pulls a piece of candy from her bag, what is the probability of her pulling out a piece of gum?

A. $\frac{1}{5}$

B. $\frac{6}{25}$

C. $\frac{9}{25}$

D. $\frac{12}{25}$

16) What is the lowest common multiple of 21 and 42?

A. 31

B. 42

C. 21

D. 882

Common Core Subject Test Mathematics Grade 6

17) Which statement is true about all squares?

   A. Both diagonals have equal measure.

   B. All sides are congruent.

   C. Both diagonals are perpendicular.

   D. All the statements are true.

18) The area of a rectangular yard is 117 square meters. What is its width if its length is 13 meters?

   A. 12 meters

   B. 15 meters

   C. 11 meters

   D. 9 meters

19) Which of the following lists shows the fractions in order from least to greatest?

$$\frac{7}{9}, \frac{5}{8}, \frac{28}{9}, \frac{21}{19}$$

   A. $\frac{21}{19}, \frac{5}{8}, \frac{7}{9}, \frac{28}{9}$

   B. $\frac{5}{8}, \frac{21}{19}, \frac{28}{9}, \frac{7}{9}$

   C. $\frac{5}{8}, \frac{7}{9}, \frac{21}{19}, \frac{28}{9}$

   D. $\frac{21}{19}, \frac{5}{8}, \frac{28}{9}, \frac{7}{9}$

Common Core Subject Test Mathematics Grade 6

20) Which statement about 6 multiplied by $\frac{8}{7}$ must be true?

   A. The product is between 4 and 5.6.

   B. The product is greater than 9.

   C. The product is equal to $\frac{37}{8}$.

   D. The product is between 6.1 and 8.1.

21) A car costing $800 is discounted 35%. Which of the following expressions can be used to find the selling price of the car?

   A. $(800)(0.35)$

   B. $800 - (800 \times 0.35)$

   C. $(800)(0.35)$

   D. $800 - (800 \times 0.30)$

22) What is the missing price factor of number 728?

$$728 = 2^3 \times 7^1 \times \ldots$$

Write your answer in the box below.

23) The distance between two cities is 3,405 feet. What is the distance of the two cities in yards?

   A. 1,135 yd.

   B. 1,335 yd.

   C. 684 yd.

   D. 2,532 yd.

Common Core Subject Test Mathematics Grade 6

24) 112 is equal to ...

  A. $-42 - (4 \times 21) + (5 \times 18)$

  B. $\left(\frac{21}{7} \times 52\right) + \left(\frac{105}{5}\right)$

  C. $\left(\left(\frac{19}{6} + \frac{32}{6}\right) \times 12\right) - \frac{32}{9} + \frac{122}{9}$

  D. $\frac{550}{10} + \frac{225}{5} + 2$

25) Mr. Jones saves $378 out of his monthly family income of $1,428. What fractional part of his income does Mr. Jones save?

  A. $\frac{18}{43}$

  B. $\frac{19}{35}$

  C. $\frac{9}{34}$

  D. $\frac{11}{40}$

26) If the area of the following trapezoid is equal to A, which equation represent x?

  A. $x = \frac{21}{A}$

  B. $x = \frac{A}{21}$

  C. $x = A + 21$

  D. $x = A - 21$

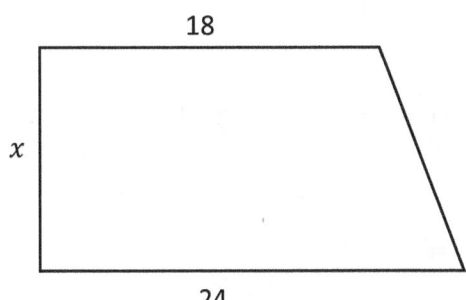

Common Core Subject Test Mathematics Grade 6

27) By what factor did the number below change from first to fifth number?

$$9, 63, 441, 3087, 21609$$

A. 7

B. 14

C. 87

D. 103

28) Based on the table below, which expression represents any value of f in term of its corresponding value of x?

A. $f(x) = 6x - \frac{1}{4}$

B. $f(x) = 5x + \frac{1}{5}$

C. $f(x) = 6x + 2\frac{1}{4}$

D. $f(x) = 7x + \frac{7}{20}$

x	0.4	1.2	2.5
$f(x)$	4.65	9.45	17.25

29) Calculate the approximate area of the following circle? (the diameter is 13)

A. 133

B. 169

C. 26

D. 531

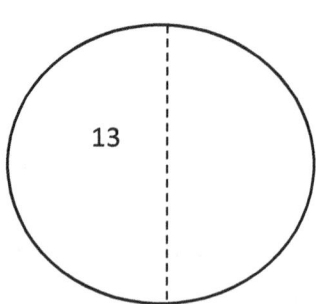

Common Core Subject Test Mathematics Grade 6

30) 97 kg = …?

   A. 97 mg

   B. 9,700 mg

   C. 970,000 mg

   D. 97,000,000 mg

31) The following graph shows the mark of seven students in mathematics. What is the mean (average) of the marks?

   A. 14

   B. 15.74

   C. 17.56

   D. 18.61

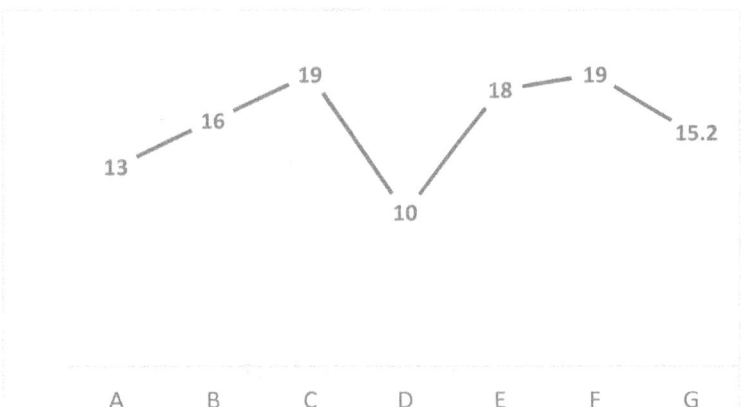

32) If the area of the following rectangular ABCD is 320, and E is the midpoint of AB, what is the area of the shaded part?

   A. 180

   B. 160

   C. 80

   D. 120

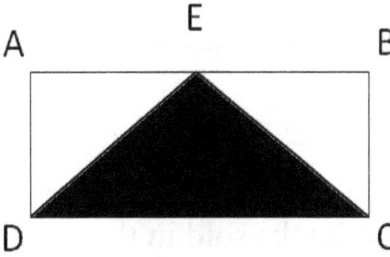

33) Which of the following statements is correct, according to the graph below?

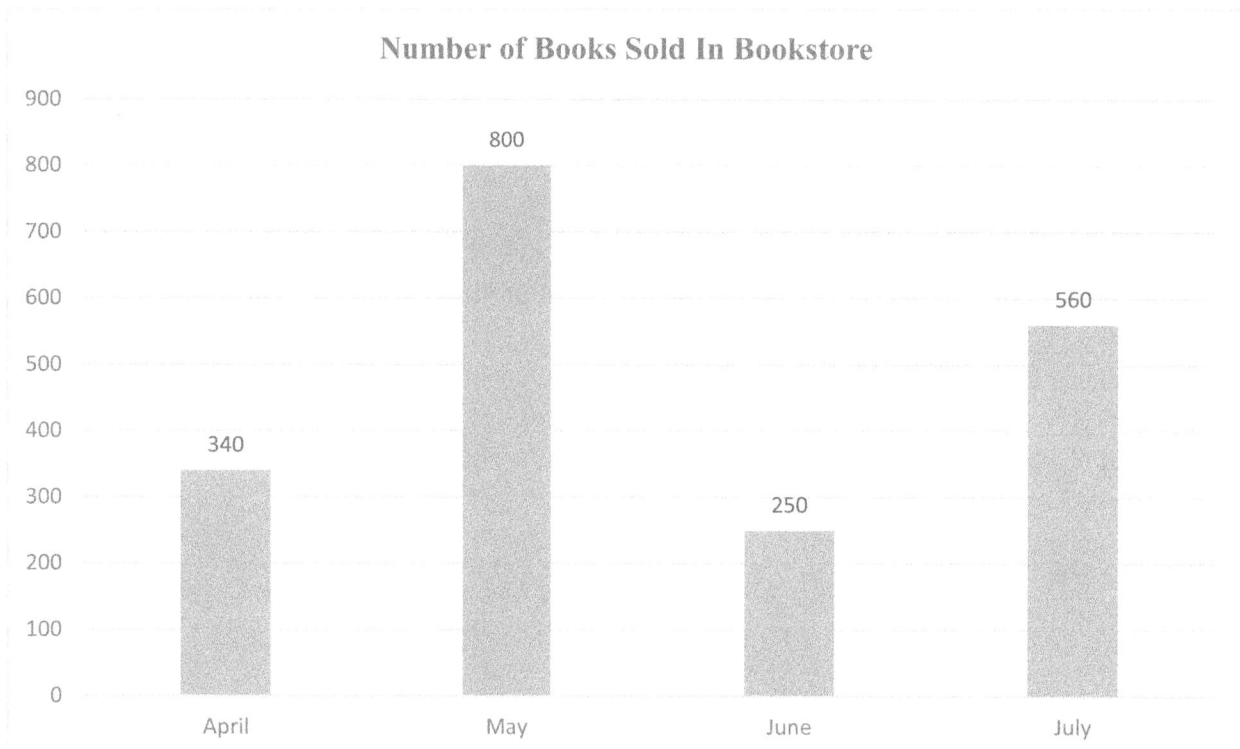

A. The number of books sold in the April was quadruple the number of books sold in the July.

B. The number of books sold in the April was less than one-fifth the number of books sold in the May.

C. The number of books sold in the June was more than one-fourth the number of books sold in the April.

D. The number of books sold in the May was equal to the number of books sold in April plus the number of books sold in the June.

Common Core Subject Test Mathematics Grade 6

34) What is the ratio between α and β $\left(\frac{\alpha}{\beta}\right)$ in the following shape?

A. $\frac{1}{15}$

B. $\frac{1}{8}$

C. $\frac{1}{30}$

D. $\frac{1}{5}$

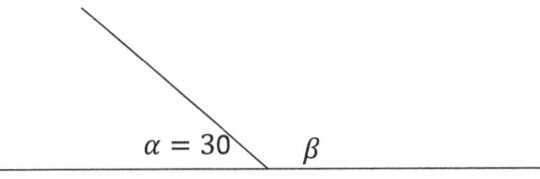

35) When point A $(-1, 8)$ is reflected over the y-axis to get the point B, what are the coordinates of point B?

A. $(-1, -8)$

B. $(1, -8)$

C. $(1, 8)$

D. $(-1, 8)$

36) In a certain bookshelf of a library, there are 45 biology books, 70 history books, and 85 language books. What is the ratio of the number of biology books to the total number of books in this bookshelf?

A. $\frac{7}{20}$

B. $\frac{1}{20}$

C. $\frac{9}{40}$

D. $\frac{3}{50}$

Common Core Subject Test Mathematics Grade 6

37) Which of the following is the correct statement?

    A. $\frac{7}{8} < 0.7$

    B. $40\% = \frac{1}{9}$

    C. $7 < \frac{32}{6}$

    D. $\frac{9}{8} > 0.93$

38) Daniel is 76 years old, quadruple as old as Henry. How old is Henry?

    A. 30 years' old

    B. 15 years' old

    C. 19 years' old

    D. 16 years' old

39) An integer is chosen at random from 5 to 25. Find the probability of not selecting a composite number?

    A. $\frac{1}{25}$

    B. $\frac{7}{10}$

    C. $\frac{7}{20}$

    D. $\frac{9}{20}$

40) Which of the following statements can be used for the following inequality?

$$\frac{x}{3} \leq 23$$

A. Sara placed x pens among 23 friends and each friend received fewer than 3 pens.

B. Sara placed 3 pens among x friends and each friend received at most 23 pens.

C. Sara placed x pens among 3 friends and each friend received fewer than 23 pens.

D. Sara placed x pens among 3 friends and each friend received at most 23 pens.

Common Core Subject Test Mathematics Grade 6

Common Core Practice Test 2

Mathematics

GRADE 6

Released *Month Year*

Common Core Subject Test Mathematics Grade 6

1) If $x = -4$, which of the following equations is true?

   A. $3x(x - 1) = 58$

   B. $7x(-2x - 5) = -84$

   C. $4(-x^2 + 7) = 38$

   D. $x(11 - x^2) = 76$

2) What is the missing prime factor of number 2,695?

   $$2,695 = 5^1 \times 7^2 \times \ldots$$

   Write your answer in the box below.

3) What is the perimeter of the following shape? (it's a right triangle)

   A. 88 cm

   B. 84 cm

   C. 64 cm

   D. 46 cm

   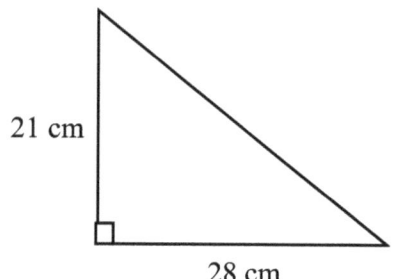

4) 40 is what percent of 16?

   A. 440 %

   B. 140 %

   C. 250 %

   D. 460 %

Common Core Subject Test Mathematics Grade 6

5) Which of the following expressions has a value of -18?

   A. $-8 + (-16 \div 8) + \frac{-7}{9} \times 9$

   B. $5 \times (-6) + (-2) \times 7$

   C. $(-8) + 15 \times 4 \div (-6)$

   D. $(-8) \times (-4) + 14$

6) 444 inches equal to …?

   A. 37 ft.

   B. 17 ft.

   C. 44 ft.

   D. 51 ft.

7) Which of the following equations is true?

   A. $0.09 = \frac{9}{10}$

   B. $\frac{44}{11} = 0.4$

   C. $3.1 = \frac{31}{10}$

   D. $\frac{45}{100} = 4.5$

Common Core Subject Test Mathematics Grade 6

8) What is the greatest common factor of 33 and 99?

   A. 33

   B. 66

   C. 99

   D. 9

9) Based on the table below, which of the following expressions represents any value of f in term of its corresponding value of x?

   A. $f(x) = 2x - \frac{1}{5}$

   B. $f(x) = 4x - 2\frac{1}{2}$

   C. $f(x) = 4x + 1$

   D. $f(x) = 2x + \frac{1}{5}$

x	2	2.5	3
$f(x)$	5.5	7.5	9.5

10) A football team won exactly 25% of the games it played during last session. Which of the following could be the total number of games the team played last season?

    A. 50

    B. 60

    C. 70

    D. 30

Common Core Subject Test Mathematics Grade 6

11) Which list shows the integer numbers listed in order from least to greatest?

   A. $-43, -31, -16, 3, 18, 28$

   B. $-16, -31, -43, 3, 18, 28$

   C. $-43, -16, -31, 3, 28, 18$

   D. $-31, -43, -16, 3, 28, 18$

12) 7 less than six times a positive integer is 41. What is the integer?

   A. 18

   B. 12

   C. 6

   D. 8

13) There are 57 blue marbles and 209 red marbles. We want to place these marbles in some boxes so that there is the same number of red marbles in each box and the same number of blue marbles in each of the boxes. How many boxes do we need?

   A. 19

   B. 11

   C. 3

   D. 23

Common Core Subject Test Mathematics Grade 6

14) The perimeter of the trapezoid below is 27. What is its area?

A. 54 cm²

B. 40 cm²

C. 60 cm²

D. 84 cm²

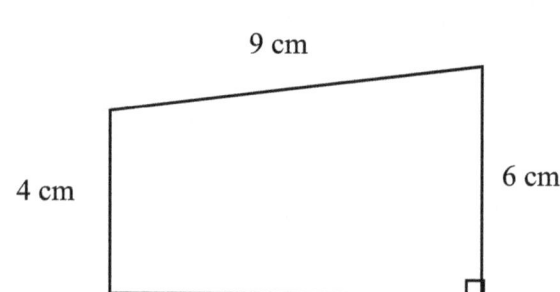

15) What is the value of the following expression?

$$798 \div 133$$

A. 12

B. 9

C. 4

D. 6

16) Solve the following equation.

$$147 = 98 + x$$

A. $x = -45$

B. $x = 45$

C. $x = -49$

D. $x = 49$

Common Core Subject Test Mathematics Grade 6

17) Car A travels 138.3 km at a given time, while car B travels 6 times the distance car A travels at the same time. What is the distance car B travels during that time?

   A. 483.8 km

   B. 846.8 km

   C. 489.4 km

   D. 829.8 km

18) Which of the following expressions has the greatest value?

   A. $4^4 - 12^2$

   B. $3^4 - 4^3$

   C. $7^3 - 4^4$

   D. $5^3 - 2^6$

19) Elise has x apples. Alvin has 51 apples, which is 21 apples less than number of apples Elise owns. If Baron has $\frac{1}{9}$ times as many apples as Elise has. How many apples does Baron have?

   A. 18

   B. 8

   C. 16

   D. 9

Common Core Subject Test Mathematics Grade 6

20) The diameter of a circle is 15π. What is the area of the circle?

   A. $15\pi^2$

   B. $\frac{15\pi^2}{2}$

   C. $\frac{225\pi^3}{4}$

   D. $\frac{115\pi^3}{4}$

21) Find the perimeter of shape in the following figure? (all angles are right angles)

   A. 54

   B. 50

   C. 48

   D. 46

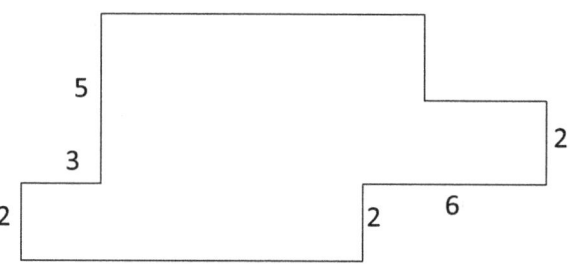

22) In the following triangle find α.

   A. 52°

   B. 50°

   C. 55°

   D. 62°

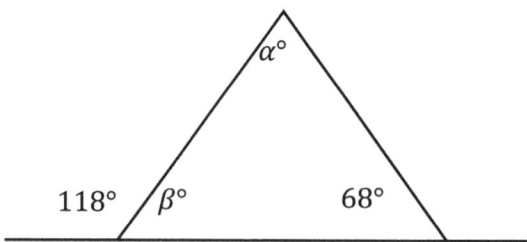

23) The price of a laptop is decreased by 16% to $672. What is its original price?

   A. $84

   B. $400

   C. $800

   D. $1,100

Common Core Subject Test Mathematics Grade 6

24) What is the probability of choosing a month starts with D in a year?

   A. $\frac{1}{3}$

   B. 1

   C. $\frac{1}{6}$

   D. $\frac{1}{12}$

25) $4(1.85) - 2.55 = \cdots?$

   A. 4.30

   B. 2.30

   C. 4.85

   D. 2.65

26) What are the values of mode and median in the following set of numbers?

   $$6, 4, 1, 8, 7, 4, 8, 9, 6, 6, 7$$

   A. Mode: 6 Median: 8

   B. Mode: 4 Median: 8

   C. Mode: 7, Median: 6

   D. Mode: 6, Median: 6

27) Which expression equivalent to $x \times 74$?

   A. $(x \times 70) + 4$

   B. $x \times 74 \times 10 \times 4$

   C. $(x \times 70) + (x \times 4)$

   D. $(x + 7) \times 10 + 4$

Common Core Subject Test Mathematics Grade 6

28) 900 mm = ...?

   A. 0.9 m

   B. 0.09 m

   C. 90 m

   D. 9,000 m

29) If point A placed at $-\frac{36}{4}$ on a number line, which of the following points has a distance equal to 21 from point A?

   A. −30

   B. 12

   C. −18

   D. A and B

30) The ratio of pens to pencils in a box is 4 to 11. If there are 180 pens and pencils in the box altogether, how many more pens should be put in the box to make the ratio of pens to pencils 1: 1?

   A. 60

   B. 132

   C. 84

   D. 48

Common Core Subject Test Mathematics Grade 6

31) Which of the following shows the numbers in increasing order?

A. $\frac{17}{7}, \frac{9}{4}, \frac{5}{8}, \frac{29}{11}$

B. $\frac{5}{8}, \frac{9}{4}, \frac{17}{7}, \frac{29}{11}$

C. $\frac{5}{8}, \frac{17}{7}, \frac{9}{4}, \frac{29}{11}$

D. $\frac{29}{11}, \frac{9}{4}, \frac{5}{8}, \frac{17}{7}$

32) What is the missing term in the given sequence?

$$12, 17, 27, 47, 87, \_\_\_, 327$$

Write your answer in the box below.

[ ]

33) If $11x - 28 = 49$, what is the value of $2x + 8$?

A. 22

B. 16

C. 7

D. 21

Common Core Subject Test Mathematics Grade 6

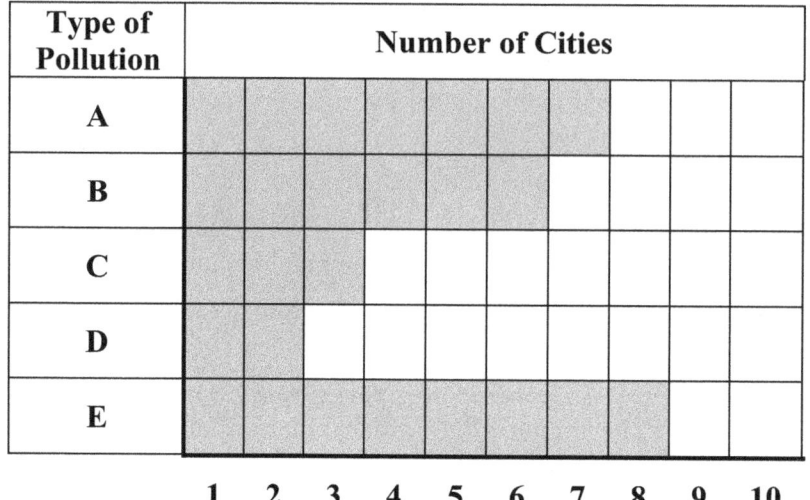

Types of air pollutions in 10 cities of a country

34) Based on the above data, what percent of cities are in the type of pollution C, B, and E respectively?

A. 70%, 30%, 80%

B. 30%, 80%, 60%

C. 30%, 60%, 80%

D. 60%, 80%, 30%

35) How many tiles of 8 cm² is needed to cover a floor of dimension 7 cm by 48 cm?

A. 42

B. 24

C. 45

D. 54

Common Core Subject Test Mathematics Grade 6

36) If there are 620 students at a school and nearly 25% of them prefer to learn Germany, approximately how many students want to learn Germany?

   A. 163

   B. 155

   C. 125

   D. 230

37) A shaft rotates 960 times in 12 seconds. How many times does it rotate in 8 seconds?

   A. 460

   B. 640

   C. 380

   D. 420

38) A card is drawn at random from a standard 52–card deck, what is the probability that the card is of the aces or queen?

   A. $\frac{2}{13}$

   B. $\frac{2}{52}$

   C. $\frac{1}{13}$

   D. $\frac{3}{52}$

Common Core Subject Test Mathematics Grade 6

39) Which of the following statement can describe the following inequality correctly? $\frac{x}{9} \geq 19$

A. David put x books in 9 shelves, and each shelf had at least 19 books.

B. David placed 9 books in x shelves so that each shelf had less than 19 books.

C. David put 19 books in x shelves and each shelf had exactly 9 books.

D. David put x books in 9 shelves, and each shelf had more than 19 books.

40) Removing which of the following numbers will change the average of the numbers to 13.8?

$$8, 10, 14, 17, 23, 7$$

A. 10

B. 14

C. 7

D. 17

Common Core Subject Test Mathematics Grade 6

Chapter 13 : Answers and Explanations

Common Core Practice Tests Answer Key

❋ Now, it's time to review your results to see where you went wrong and what areas you need to improve!

Practice Test - 1						Practice Test - 2					
1	A	16	B	31	B	1	B	16	D	31	B
2	B	17	D	32	B	2	11	17	D	32	167
3	C	18	D	33	C	3	B	18	A	33	A
4	D	19	C	34	D	4	C	19	B	34	C
5	B	20	D	35	C	5	C	20	C	35	A
6	B	21	B	36	C	6	A	21	C	36	B
7	A	22	13	37	D	7	C	22	B	37	B
8	C	23	A	38	C	8	A	23	C	38	A
9	D	24	C	39	C	9	B	24	D	39	A
10	C	25	C	40	D	10	B	25	C	40	A
11	B	26	B			11	A	26	D		
12	C	27	A			12	D	27	C		
13	B	28	C			13	A	28	A		
14	A	29	A			14	B	29	D		
15	B	30	D			15	D	30	C		

Common Core Subject Test Mathematics Grade 6

Common Core Subject Test Mathematics Grade 6

Practice Test 1
Common Core - Mathematics
Answers and Explanations

1) **Answer: A**

For one hour he earns $14, then for t hours he earns $14t$. If he wants to earn at least $165, therefor, the number of working hours multiplied by 14 must be equal to 165 or more than 165. $14t \geq 165$

2) **Answer: B**

Plug in the value of x and y and use order of operations rule. $x = 4$ and $y = -1$

$3(x - 3y) + (11 - 2x)^2 = 3(4 - 3(-1)) + (11 - 2(4))^2 = 3(4 + 3) + (3)^2 = 21 + 9 = 30$

3) **Answer: C**

$\frac{178}{7} \cong 25.43 \cong 25.4$

4) **Answer: D**

$8(7x - 11) = (8 \times 7x) - (8 \times 11) = (8 \times 7)x - (8 \times 11) = 56x - 88$

5) **Answer: B**

17% of the volume of the solution is alcohol. Let x be the volume of the solution. Then: 17% of x = 85 ml $\Rightarrow$ 0.17 x = 85 $\Rightarrow$ x = 85 ÷ 0.17 = 500

6) **Answer: B**

The coordinate plane has two axes. The vertical line is called the y-axis and the horizontal is called the x-axis. The points on the coordinate plane are address using the form (x, y). The point A is two unit on the left side of x-axis; therefore, its x value is -2 and it is two unit up, therefore its y axis is 2. The coordinate of the point is: $(-2, 2)$

7) **Answer: A**

108: 36 = 27: 9

27 × 4 = 108   And   9 × 4 = 36

Common Core Subject Test Mathematics Grade 6

8) Answer: C

Opposite number of any number x is a number that if added to x, the result is 0. Then:

$14 + (-14) = 0$ and $6 + (-6) = 0$

9) Answer: D

Solve for x. $-3 \leq 7x - 3 < 11 \Rightarrow$ (add 3 all sides) $-3 + 3 \leq 7x - 3 + 3 < 11 + 3$

$\Rightarrow 0 \leq 7x < 14 \Rightarrow$ (divide all sides by 7) $0 \leq x < 2$

x is between 0 and 2. Choice D represent this inequality.

10) Answer: C

$-18 = 41 - x$

First, subtract 41 from both sides of the equation. Then:

$-18 - 41 = 41 - 41 - x \rightarrow -59 = -x$

Multiply both sides by $(-1) \rightarrow x = 59$

11) Answer: B

The ratio of boy to girls is 3:4. Therefore, there are 3 boys out of 7 students. To find the answer, first divide the total number of students by 7, then multiply the result by 3

$154 \div 7 = 22 \Rightarrow 22 \times 3 = 66$

12) Answer: C

$(64 + 8) \div (24) = (72) \div (24)$

The prime factorization of 72 is: $2 \times 2 \times 2 \times 3 \times 3$

The prime factorization of 24 is: $2 \times 2 \times 2 \times 3$

Therefore: $(72) \div (24) = (2 \times 2 \times 2 \times 3 \times 3) \div (2 \times 2 \times 2 \times 3)$

13) Answer: B

The slope of the line is: $\frac{y_2 - y_1}{x_2 - x_1} = \frac{8-3}{5-4} = \frac{5}{1} = 5$

The equation of a line can be written as:

$y - y_0 = m(x - x_0) \rightarrow y - 3 = 5(x - 4) \rightarrow y - 3 = 5x - 20 \rightarrow y = 5x - 17$

14) Answer: A

Volume of a box = length × width × height = $11 \times 4 \times 3 = 132$

WWW.MathNotion.Com

Common Core Subject Test Mathematics Grade 6

15) Answer: B

Probability = $\frac{number\ of\ desired\ outcomes}{number\ of\ total\ outcomes} = \frac{12}{15+6+12+17} = \frac{12}{50} = \frac{6}{25}$

16) Answer: B

Prime factorizing of $21 = 3 \times 7$

Prime factorizing of $42 = 2 \times 3 \times 7$

LCM= $2 \times 3 \times 7 = 42$

17) Answer: D

In any square, all the statements are true.

18) Answer: D

Let y be the width of the rectangle. Then; $13 \times y = 117 \rightarrow y = \frac{117}{13} = 9$

19) Answer: C

Let's compare each fraction: $\frac{5}{8} < \frac{7}{9} < \frac{21}{19} < \frac{28}{9}$

Only choice C provides the right order.

20) Answer: D

$6 \times \frac{8}{7} = \frac{48}{7} = 6.86$

A. $6.86 > 5.6$

B. $6.86 < 9$

C. $\frac{37}{8} = 4.625 \neq 6.86$

D. $6.1 < 6.86 < 8.1$ This is the answer!

21) Answer: B

To find the discount, multiply the number $(100\% -$ rate of discount$)$

Therefore; $800(100\% - 35\%) = 800(1 - 0.35) = 800 - (800 \times 0.35)$

22) Answer: 13.

$728 = 2^3 \times 7^1 \times 13^1$

23) Answer: A

1 yard = 3 feet

Therefore, 3,405 ft.$\times \frac{1\ yd}{3\ ft} = 1,135$ yd

Common Core Subject Test Mathematics Grade 6

24) Answer: C

Simplify each option provided.

A. $-42 - (4 \times 21) + (5 \times 18) = -42 - 84 + 90 = -36$

B. $\left(\frac{21}{7} \times 52\right) + \left(\frac{105}{5}\right) = 156 + 21 = 177$

C. $\left(\left(\frac{19}{6} + \frac{32}{6}\right) \times 12\right) - \frac{32}{9} + \frac{122}{9} = \left(\left(\frac{19+32}{6}\right) \times 12\right) - \frac{32}{9} + \frac{122}{9} = \left(\left(\frac{51}{6}\right) \times 12\right) + \frac{122-32}{9} = (51 \times 2) + \frac{90}{9} = 102 + 10 = 112$ (this is the answer)

D. $\frac{550}{10} + \frac{225}{5} + 2 = \frac{550+450}{10} + 2 = 100 + 2 = 102$

25) Answer: C

378 out of 1,428 equals to $\frac{378}{1,428} = \frac{9}{34}$

26) Answer: B

The area of the trapezoid is: $area = \frac{(base\ 1 + base\ 2)}{2} \times height = \left(\frac{18+24}{2}\right)x = A \rightarrow$

$21x = A \rightarrow x = \frac{A}{21}$

27) Answer: A

$\frac{63}{9} = 7, \frac{441}{63} = 7, \frac{3,087}{441} = 7, \frac{21,609}{3,087} = 7$

Therefore, the factor is 7.

28) Answer: C

Plug in the value of x into the function $f(x)$. First, plug in 0.4 for x.

A. $f(x) = 6x - \frac{1}{4} = 6(0.4) - \frac{1}{4} = 2.15 \neq 4.65$

B. $f(x) = 5x + \frac{1}{5} = 5(0.4) + \frac{1}{5} = 2.2 \neq 4.65$

C. $f(x) = 6x + 2\frac{1}{4} = 6(0.4) + 2\frac{1}{4} = 2.4 + 2.25 = 4.65$ This is correct!

Plug in other values of x. $x = 1.2$

$f(x) = 6x + 2\frac{1}{4} = 6(1.2) + 2.25 = 9.45$ This one is also correct. $x = 2.5$

$f(x) = 6x + 2\frac{1}{4} = 6(2.5) + 2.25 = 17.25$ This one works too!

D. $f(x) = 7x + \frac{7}{20} = 7(0.4) + \frac{7}{20} = 3.15 \neq 4.65$

Common Core Subject Test Mathematics Grade 6

29) Answer: A

The diameter of a circle is twice the radius. Radius of the circle is $\frac{13}{2}$.

Area of a circle $= \pi r^2 = \pi(\frac{13}{2})^2 = 42.25\pi = 42.25 \times 3.14 = 132.67 \cong 133$

30) Answer: D

1 kg= 1,000 g and 1 g = 1,000 mg

97 kg= 97× 1,000 g =97 × 1,000 × 1,000 = 97,000,000 mg

31) Answer: B

Average (mean) $= \frac{\text{sum of terms}}{\text{number of terms}} = \frac{13+16+19+10+18+19+15.2}{7} = 15.74$

32) Answer: B

Since, E is the midpoint of AB, then the area of all triangles DAE, DEF, CFE and CBE are equal.

Let x be the area of one of the triangles, then:

$4x = 320 \rightarrow x = 80$

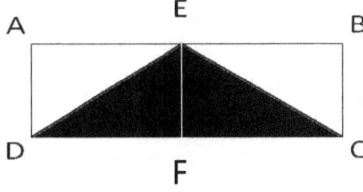

The area of DEC $= 2x = 2(80) = 160$

33) Answer: C

A. Number of books sold in April is: 340

Number of books sold in July is: 560 $\rightarrow \frac{340}{560} = \frac{17}{28} \neq 4$

B. Number of books sold in April is: 340

One-fifth the number of books sold in May is: $\frac{800}{5} = 160 \rightarrow 340 > 160$

C. Number of books sold in June is: 250

one-fourth the number of books sold in April is: $\frac{340}{4} = 85 \rightarrow 250 > 85$ (correct)

D. $340 + 250 = 590 > 800$

34) Answer: D

α and β are supplementary angles. The sum of supplementary angles is 180 degrees.

$\alpha + \beta = 180° \rightarrow \beta = 180° - \alpha = 180° - 30° = 150°$; Then, $\frac{\alpha}{\beta} = \frac{30}{150} = \frac{1}{5}$

Common Core Subject Test Mathematics Grade 6

35) Answer: C

When points are reflected over y-axis, the value of y in the coordinates doesn't change and the sign of x changes. Therefore, the coordinates of point B is $(1, 8)$.

36) Answer: C

Number of biology book: 45

Total number of books; $45 + 70 + 85 = 200$

The ratio of the number of biology books to the total number of books is: $\frac{45}{200} = \frac{9}{40}$

37) Answer: D

A. $\frac{7}{8} < 0.7$ $\frac{7}{8} = 0.875$. Therefore, this inequality is not correct.

B. $40\% = \frac{1}{9}$ $40\% = \frac{2}{5}$, not $\frac{1}{9}$.

C. $7 < \frac{32}{6}$ $\frac{32}{6} = 5.33$. Therefore, this inequality is not correct.

D. $\frac{9}{8} > 0.93$ $\frac{9}{8} = 1.125 \to 1.125 > 0.93$, this inequality is correct.

38) Answer: C

Henry is x years old, then, $4x = 76 \to x = \frac{76}{4} = 19$

39) Answer: C

There are 20 integers from 5 to 25. Set of numbers that are not composite between 5 and 25 is: A= {5, 7, 11, 13, 17, 19, 23}

7 integers are not composite. Probability of not selecting a composite number is:

Probability $= \frac{number\ of\ desired\ outcomes}{number\ of\ total\ outcomes} = \frac{7}{20}$

40) Answer: D

Let's write the inequality for each statement.

A. $\frac{x}{23} < 3$

B. $\frac{3}{x} \leq 23$

C. $\frac{x}{3} < 23$

D. $\frac{x}{3} \leq 23$ This is the inequality provided in the question.

Common Core Subject Test Mathematics Grade 6

Practice Test 2
Common Core - Mathematics
Answers and Explanations

1) Answer: B

Plugin the value of x in the equations. $x = -4$, then:

A. $3x(x - 1) = 58 \to 3(-4)(-4 - 1) = (-12)(-5) = 60 \neq 58$

B. $7x(-2x - 5) = -84 \to 7(-4)(-2(-4) - 5) = -28(8 - 5) = -84 = -84$

C. $4(-x^2 + 7) = 38 \to 4\left(-(-4)^2 + 7\right) = 4(-16 + 7) = 4(-9) = -36 \neq 38$

D. $x(11 - x^2) = 76 \to (-4)(11 - (-4)^2) = (-4)(11 - 16) = (-4)(-5) = 20 \neq 76$

2) Answer: 11.

Let x be the missing prime factor of 2,695.

$2,695 = 5 \times 7 \times 7 \times x \Rightarrow x = \frac{2,695}{245} \Rightarrow x = 11$

The missing prime factor of 2,695 is 11.

3) Answer: B

Use Pythagorean theorem to find the hypotenuse of the triangle.

$a^2 + b^2 = c^2 \to 21^2 + 28^2 = c^2 \to 441 + 784 = c^2 \to 1,225 = c^2 \to c = 35$

The perimeter of the triangle is: $21 + 28 + 35 = 84$

4) Answer: C

Use percent formula: Part $= \frac{percent}{100} \times$ whole.

$40 = \frac{percent}{100} \times 16 \Rightarrow 40 = \frac{percent \times 16}{100} \Rightarrow 40 = \frac{percent \times 4}{25}$, multiply both sides by 25.

$1,000 = percent \times 4$, divide both sides by 4.

$250 = percent$; The answer is 250%

5) Answer: C

Let's check the options provided.

A. $-8 + (-16 \div 8) + \frac{-7}{9} \times 9 \to -8 + (-2) + (-7) = -17$

WWW.MathNotion.Com

Common Core Subject Test Mathematics Grade 6

B. $5 \times (-6) + (-2) \times 7 = (-30) + (-14) = -44$

C. $(-8) + 15 \times 4 \div (-6) = -8 + 60 \div (-6) = -8 - 10 = -18$

D. $(-8) \times (-4) + 14 = 32 + 14 = 46$

6) Answer: A

1 feet = 12 inches. Then: $444 \text{ in} \times \frac{1 \text{ ft}}{12 \text{ in}} = \frac{444}{12} \text{ ft} = 37 \text{ft}$

7) Answer: C

A. $0.09 = \frac{9}{100}$

B. $\frac{44}{11} = 4$

C. $3.1 = 3\frac{1}{10} = \frac{31}{10}$

D. $\frac{45}{100} = 0.45$

8) Answer: A

Prime factorizing of $33 = 3 \times 11$

Prime factorizing of $99 = 3 \times 3 \times 11$

To find Greatest Common Factor, multiply the common factors of both numbers.

GCF = $3 \times 11 = 33$

9) Answer: B

Plug in the values of x in the equations provided.

A. $f(x) = 2x - \frac{1}{5} = 2(2) - \frac{1}{5} = 4 + 0.2 = 3.8 \neq 5.5$

B. $f(x) = 4x - 2\frac{1}{2} = 4(2) - \frac{5}{2} = 5.5$

C. $f(x) = 4x + 1 = 4(2) + 1 = 9 \neq 5.5$

D. $f(x) = 2x + \frac{1}{5} = 2(2) + \frac{1}{5} = 4.2 \neq 5.5$

10) Answer: B

Choices A, C and D are incorrect because 25% of each of the numbers is a non-whole number.

A. 50, 25% of $50 = 0.25 \times 50 = 12.5$

B. 60, 25% of $60 = 0.25 \times 60 = 15$

Common Core Subject Test Mathematics Grade 6

C. 70, $25\% \ of \ 70 = 0.25 \times 70 = 17.5$

D. 30, $25\% \ of \ 30 = 0.25 \times 30 = 7.5$

11) **Answer: A**

$-43 < -31 < -16 < 3 < 18 < 28$

Then choice A

12) **Answer: D**

Let x be the integer. Then: $6x - 7 = 41$

Add 7 both sides: $6x = 48$

Divide both sides by 6: $x = 8$

13) **Answer: A**

First, we need to find the GCF (Greatest Common Factor) of 209 and 57.

$209 = 11 \times 19$

$57 = 3 \times 19 \rightarrow$ GFC = 19; Therefore, we need 19 boxes.

14) **Answer: B**

The perimeter of the trapezoid is 27.

Therefore, the missing side (height) is $= 27 - 4 - 6 - 9 = 8$

Area of the trapezoid: $A = \frac{1}{2}h(b1 + b2) = \frac{1}{2}(8)(4+6) = 40$

15) **Answer: D**

$798 \div 133 = \frac{798}{133} = \frac{114}{19} = 6$

16) **Answer: D**

$147 = 98 + x$

Subtract 98 from both sides of the equation. Then:

$x = 147 - 98 = 49$

17) **Answer: D**

Distance that car B travels $= 6 \times$ distance that car A travels

$= 6 \times 138.3 = 829.8$ Km

18) **Answer: A**

A. $4^4 - 12^2 = 256 - 144 = 112$

Common Core Subject Test Mathematics Grade 6

B. $3^4 - 4^3 = 81 - 64 = 17$

C. $7^3 - 4^4 = 343 - 256 = 87$

D. $5^3 - 2^6 = 125 - 64 = 61$

19) Answer: B

Elise has x apple which is 21 apples more than number of apples Alvin owns.

Therefore: $x - 21 = 51 \rightarrow x = 51 + 21 = 72$

Elise has 72 apples.

Let y be the number of apples that Baron has. Then: $y = \frac{1}{9} \times 72 = 8$

20) Answer: C

The radius of the circle is: $\frac{15\pi}{2}$

The area of circle: $\pi r^2 = \pi(\frac{15\pi}{2})^2 = \pi \times \frac{225\pi^2}{4} = \frac{225\pi^3}{4}$

21) Answer: C

Let x and y be two sides of the shape. Then:

$x + 2 + 2 = 5 + 2 \rightarrow x = 3$

$y + 9 + 3 = 11 + 6 \rightarrow y + 12 = 17 \rightarrow y = 5$

Then, the perimeter is:

$2 + 11 + 2 + 6 + 2 + 5 + 3 + 9 + 5 + 3 = 48$

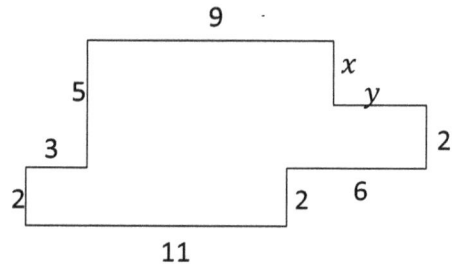

22) Answer: B

Supplementary angles add up to 180 degrees.

$\beta + 118° = 180° \rightarrow \beta = 180° - 118° = 62°$

The sum of all angles in a triangle is 180 degrees. Then:

$\alpha + \beta + 68° = 180° \rightarrow \alpha + 62° + 68° = 180°$

$\rightarrow \alpha + 130° = 180° \rightarrow \alpha = 180° - 130° = 50°$

23) Answer: C

Let x be the original price.

If the price of a laptop is decreased by 16% to $672, then: 84% of $x = 672$

$\Rightarrow 0.84x = 672 \Rightarrow x = 672 \div 0.84 = 800$

Common Core Subject Test Mathematics Grade 6

24) Answer: D

One months, December, in 12 months start with D, then:

Probability = $\frac{number\ of\ desired\ outcomes}{number\ of\ total\ outcomes} = \frac{1}{12}$

25) Answer: C

$4(1.85) - 2.55 = 7.4 - 2.55 = 4.85$

26) Answer: D

First, put the numbers in order from least to greatest: 1, 4, 4, 6, 6, 6, 7, 7, 8, 8, 9.

The Mode of the set of numbers is: 6 (the most frequent numbers)

Median is: 6 (the number in the middle)

27) Answer: C

$x \times 74 = x \times (70 + 4) = (x \times 70) + (x \times 4)$

28) Answer: A

$1\ m = 1,000\ mm$

$1\ mm = 0.001\ m$

Then, $900\ mm = 900 \times 0.001\ m = 0.9\ m$

29) Answer: D

If the value of point A is greater than the value of point B, then the distance of two points on the number line is: value of A− value of B

A. $-\frac{36}{4} - (-30) = -9 + 30 = 21 = 21$

B. $12 - \left(-\frac{36}{4}\right) = 12 + 9 = 21 = 21$

C. $-18 - \left(-\frac{36}{4}\right) = -18 + 9 = -9 \neq 21$

30) Answer: C

The ratio of pens to pencils is 4: 11. Therefore there are 4 pens out of all 15 pens and pencils. To find the answer, first dived 95 by 15 then multiply the result by 4.

$180 \div 15 = 12 \rightarrow 12 \times 4 = 48$

There are 48 pens and 132 pencils (180−48). Therefore, 84 more pens should be put in the box to make the ratio 1: 1.

Common Core Subject Test Mathematics Grade 6

31) Answer: B

$\frac{29}{11} \cong 2.64$ $\frac{9}{4} = 2.25$ $\frac{5}{8} = 0.625$ $\frac{17}{7} \cong 2.43$

Then: $\frac{5}{8} < \frac{9}{4} < \frac{17}{7} < \frac{29}{11}$

32) Answer: 167

Find the difference of each pairs of numbers:

12, 17, 27, 47, 87, ___, 327

The difference of 12 and 17 is 5, 17 and 27 is 10, 27 and 47 is 20, 47 and 87 is 40, 87 and next number should be 80. The number is 87 + 80 = 167

33) Answer: A

$11x - 28 = 49 \to 11x = 49 + 28 = 77 \to x = \frac{77}{11} = 7$

Then, $2x + 8 = 2(7) + 8 = 14 + 8 = 22$

34) Answer: C

Percent of cities in the type of pollution C: $\frac{3}{10} \times 100 = 30\%$

Percent of cities in the type of pollution B: $\frac{6}{10} \times 100 = 60\%$

Percent of cities in the type of pollution E: $\frac{8}{10} \times 100 = 80\%$

35) Answer: A

The area of the floor is: 7 cm × 48 cm = 336 cm

The number of tiles needed = 336 ÷ 8 = 42

36) Answer: B

Number of students prefer to learn Germany= $25\% \, of \, 620 = \frac{25}{100} \times 620 = 155$

37) Answer: B

The shaft rotates 960 times in 12 seconds. Then, the number of rotates in 8 second equals to:

$\frac{960 \times 8}{12} = 640$

38) Answer: A

The probability of choosing a aces or queen is $\frac{4}{52} + \frac{4}{52} = \frac{2}{13}$

WWW.MathNotion.Com

Common Core Subject Test Mathematics Grade 6

39) Answer: A

Let's write an inequality for each statement.

A. $\frac{x}{9} \geq 19$ (this is the same as the inequality provided)

B. $\frac{9}{x} < 19$

C. $\frac{19}{x} = 9$

D. $\frac{x}{9} > 19$

40) Answer: A

Check each option provided:

A. 10 $\frac{8+14+17+23+7}{5} = \frac{69}{5} = 13.8$

B. 14 $\frac{8+10+17+23+7}{5} = \frac{65}{5} = 13$

C. 7 $\frac{8+10+14+17+23}{5} = \frac{72}{5} = 14.4$

D. 17 $\frac{8+10+14+23+7}{5} = \frac{62}{5} = 12.4$

"End"

www.ingramcontent.com/pod-product-compliance
Lightning Source LLC
Chambersburg PA
CBHW080439110426
42743CB00016B/3217